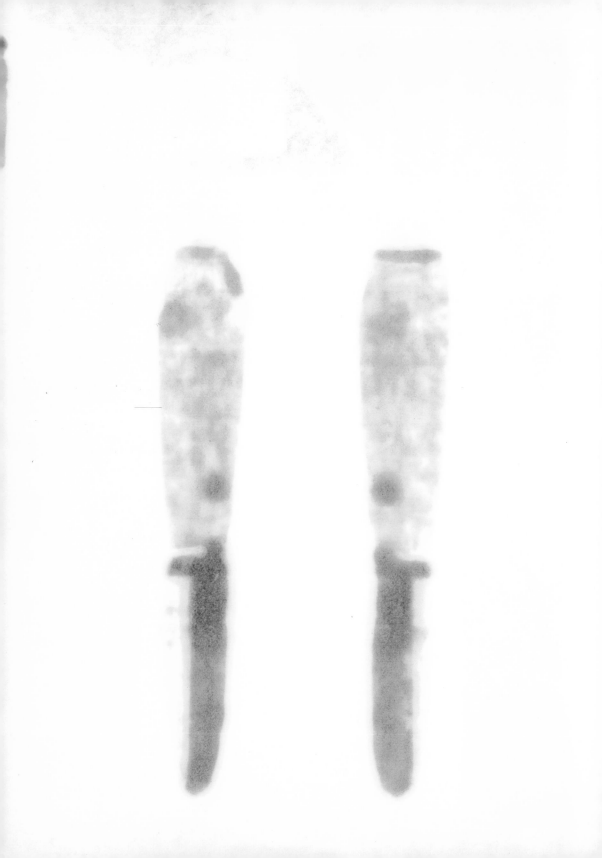

Close to 500 bhp is available to the Alpine-Renault A442s from their turbocharged Group 6 2-litre V6 engines

Turbocharging
and
Supercharging
for maximum power *and* torque

LJK Setright

Foulis

ISBN 0 85429 184 9

First published September 1976

a **FOULIS MOTORING book**

J H Haynes and Company Limited

Sparkford Yeovil Somerset BA22 7JJ

Printed by J H Haynes and Company Limited Sparkford Yeovil Somerset BA22 7JJ
Bound by The Western Book Company Limited Maesteg Glamorgan S Wales

Editor Tim Parker
Jacket design Edward Piper
Illustrator Terry Davey

Contents

Credits

Grateful thanks are extended to all those who have helped with this book, both knowingly and unknowingly. Apologies are given to those whose anonymity does not permit us to mention them.

AiResearch Industrial Division of Garratt Corporation, Alfa Romeo SpA, BMW AG, Les Brazier, Broadspeed Engineering Ltd, Castrol Limited, Daimler-Benz AG, The Ford Motor Company, General Motors, Holset Engineering Co. Ltd, The Institution of Mechanical Engineers, London Art Technical Drawings Limited, OM SpA, Adam Opel AG, Dr Ing h.c. F. Porsche AG, Dennis Priddle, Rajay Industries Inc, Turbo May GmbH + Co KG, and TVR Engineering Ltd.

000 bhp at 8,000 rpm
upercharge for this sort of power

Chapter 1
Unnatural aspiration

When the inlet valves of any normally aspirated engine open to admit a fresh charge of air (together with fuel, unless that be introduced to the cylinder by direct injection) into the cylinders, it is left to God to ensure that the air does actually find its way. This provision of an ambient *atmosphere* at a sea level pressure of about 1 kg per square centimetre (1 atm) is not always sufficient for the task: so although the engine may breathe reasonably deeply, to inhale perhaps 80% of the volume displaced by the piston descending on its induction or suction stroke, at higher engine speeds there will be symptoms of asthma as the time available for the cylinder to draw in a lungful becomes shorter. There are recognized ways of overcoming this difficulty, ways that are well known to the motor sporting fraternity: by the use of inlet and exhaust pipes tuned to resonate at a chosen high frequency, and in conjunction with valve opening periods of extravagant duration, it is possible so to exploit nature's susceptibilities as to achieve a breathing efficiency of 100% or even more at the chosen speed — but at all speeds significantly lower or higher, the breathing of the engine becomes very inefficient and unsatisfactory, sometimes to the point where the engine will not run at all at any speed less than about one half of that chosen for maximum efficiency. Should you wish your engine to retain the flexibility of the mild-mannered tourer, while yet enjoying as much power as (or even more than) the highly tuned racing engine, then instead of relying upon God and His atmosphere, you must go to the Devil and be prepared to take the consequences: instead of relying on natural aspiration from the ambient atmosphere you must have recourse to mechanically forced induction that creates an artificial high-pressure atmosphere.

11

Turbocharging and Supercharging for Maximum Power and Torque

It will involve taking some new risks — characteristically concerned with heat — but in exchange you will come to accept a new freedom from certain constraints. Supercharging, whatever the means of doing it, is a device to increase the mass flow of air into the engine, and thus increase the engine's capacity for doing work, which is dependent upon the rate at which fuel can be burned in that air. It is a fine way of making an engine more powerful, and it is probably safe to state that most people in search of power have at one time or another had to have recourse to the supercharger — at least, since about 1902. This is because of the considerable thermal problems associated with supercharging, and inseparable from the fact that as the mass flow of air into and through the engine is increased, the heat flow is increased in at least equal measure. This makes it all the more intriguing that in 1901 Sir Dugald Clerk (the inventor of a two-stroke engine) was advocating the use of supercharging to create an artificially dense atmosphere from which the engine could breathe, the extra air being meant to control the maximum temperature reached in the combustion phase — but in the process he found a power increase of 6%, and so the search for supercharged power began. In 1902 Louis Renault had ideas for it, in 1909 Lee Chadwick made it work effectively, and in the same year Büchi in Switzerland hit on the idea of a supercharger driven by an exhaust turbine for diesel engines. After that, developments were rapid: first on aero engines during the Great War (because the supercharger or turbocharger could compensate for the low atmospheric density at high altitudes) and then in the 1920s for sports and racing cars, most of the pioneering being done by Mercédès and Fiat. From then on, supercharging for racing cars never looked back, until in 1951 Aurelio Lampredi turned the tables on the hitherto unbeatable supercharged Alfa Romeo Grand Prix cars with his unsupercharged Ferrari which, with three times their engine displacement, was their equal in power and their superior in fuel consumption.

The power of a piston engine is the product of its piston area, the length of the piston stroke, the number of strokes per minute, and the mean effective pressure on the pistons. Every one of these factors except the last is limited by purely mechanical considerations: they apply with equal force to all engines, whether supercharged or otherwise, and can be accepted as a constant when studying how to achieve the best performance from any given engine. Thus, the maximum rotational rate of its crankshaft will be determined by such things as its vibration characteristics, bearing limitations, inertia loadings, and so on, and the same limit will apply whatever we do to the engine's breathing.

All that remains is the mean effective pressure. This is capable of being increased by supercharging, but the increase may not be proportional to the degree of supercharge, for reasons to which we shall come later — and the difference will represent a loss in operating efficiency that can be lumped together with the frictional losses that are automatically taken into account in calculating the mean effective pressure from a knowledge of the other factors determining the power output.

Whether by supercharging or by any of the tricks current in the unblown engine tuner's trade, to increase the mean effective pressure is equivalent to increasing the volumetric efficiency of the engine: in other words the mean effective pressure is related to how well the engine breathes its charge of fuel and air.

Within the limits imposed by the mechanical features of the engine, this volumetric efficiency can be increased without recourse to forced induction, by resorting to such methods as tuned inlet and exhaust pipes, high compression ratios, increased lift, duration and overlap of the valves, modified ignition timing, and so on. Provided that these processes are not carried too far, the effect can be wholly beneficial, not only to power, but also to economy (in the sense of improved specific fuel consumption, which is measured in quantity of fuel per horsepower per hour) and even to overall flexibility, since an increase in the usefulness of the top end of the rev range is as valuable as an increase at the bottom. To arrive at such a specification will cost money, so it cannot be thought of as altogether something for nothing; nevertheless if the tuners' greed for power be kept under tight rein, it is possible to effect an all-round improvement in the behaviour of the engine by such means. Demand more, however, and the stage is set for the conversion of what was once a well-mannered engine into something fierce, mettlesome and intransigent, producing plenty of power inside a certain limited rev range, and being more of a liability than an asset outside it.

Instead of all these devices which seek to make the most of what the natural ambient atmosphere can offer, one can supply the engine with an artificially intensified atmosphere, by mechanical means, simply by pumping in more air than nature is content to do. In other words, one raises the air pressure at the intake ports, so that when the inlet valves open, a greater mass of air enters the cylinders in the available time. Some modifications to the engine, such as to the compression ratio and ignition timing, may still be necessary; but even a modest degree of supercharge is bound to fill out the power curve in a most satisfying way, without necessarily having any detrimental effect on the engine's inherent good manners. Once again it is important not to demand too much, otherwise the pressure-charged engine can develop a temperament just as unmanageable as that of a highly tuned unblown engine, though that situation is likely to correspond to a vast increase in power output and an even more impressive rise in fuel consumption.

Which course to adopt, and how far to pursue it, are matters that must be decided by reference to the nature and duration of the duties the engine must perform. In many classes of racing, the supercharger is quite simply forbidden by the current rules, and that must be that. But there are other classes of competition where forced induction is permitted, and then the choice is more difficult to make — especially if some kind of handicapping system is involved, seeking to equate a pressure-charged engine of a given displacement with an unsupercharged engine of some arbitrarily greater displacement. In a race of long duration over a course that allows a more or less constant speed to be maintained, and makes no particular demands on accuracy and promptness of throttle response, the turbocharged engine is likely to be at a premium, as in recent years has been the case at Indianapolis. At the other extreme, sprints, and hill climbs of relatively short duration, but demanding an amplitude of power and torque, to permit acceleration through and between corners without too much time being lost on gear changing, may have their demands best met by an engine with mechanically driven supercharger, making the most of its accurate and instantaneous throttle response, and discounting its incredible thirst for fuel and its possible want of long term reliability. Between these two extremes are to be found road racers in which it is feasible to keep the engine working within a narrow speed range, by free but predictable use of a gearbox, in which case the desirability of prompt response and the importance of minimising the load of fuel to be carried (or alternatively of eliminating the danger of having to stop for refuelling) make the unsupercharged engine the most appropriate.

In all competitions of the type so far discussed, the engine will be called upon to deliver its utmost power for the major part of its operating time. Forty years ago, when there were some extremely powerful supercharged cars engaged in Grand Prix racing, it was estimated that they were only driven with wide-open throttle for about one third of a race, the rest of the time being equally divided between the overrun condition and braking. Improvements in chassis and suspension design and all ancillaries such as tyres and brakes have significantly altered things, however, and today it is likely that the car would be under impulsion from its engine for at least 80% of the time. In rallying and in ordinary high-performance road driving, the situation is very different: it is unlikely that full power will be demanded for bursts much longer than ten or fifteen seconds, and when driving on public roads there will be long periods when little more than 10% or so of the engine's potential may be realised. In these circumstances it may be very difficult indeed to decide how best to secure the performance required. It may come from a large and ponderous but lightly stressed unsupercharged engine, from a smaller, lighter, but more highly tuned and commensurately more highly stressed unsupercharged engine, or from a supercharged engine of similar size that will not be so highly stressed mechanically in its bottom half, but will be subjected to greater thermal stresses in its top half.

If the problem revolves round an existing engine that may be developed either in blown or unblown form, the latter may require a lot of expensive modification to allow it to run at higher crankshaft rpm, so as to increase the mass of air flowing through it in a given time: the supercharged version can be left unchanged in these respects, the mechanical compressor sufficing to give the same result; but the supercharger will need power to drive it which it must derive from the engine, and when run light as in gentle traffic driving, it may be mechanically inefficient as a result. When under load, it will be thermally inefficient, so in either circumstance its fuel consumption is likely to be higher than that of the highly tuned unblown version of similar power. On the other hand, the unblown engine may have to be furnished with a gearbox offering a large number of relatively closely spaced ratios, and this may introduce problems of inconvenience or capital expense that may weigh as heavily as the onerous fuel consumption of the blown engine.

Then there is the turbocharged engine to consider: it too may be of similar size to the supercharged or unblown but tuned engines, but when running light it will be mechanically more efficient than the supercharged engine, because its turbine makes no significant depredations from the engine's power output in these circumstances. Its fuel consumption may therefore be more acceptable, but the penalty to be paid is in the lack of control so far inseparable from turbocharging: it is a matter of immediacy of response to the throttle, a matter that has exercised many development engineers in the past few years, without any of them achieving altogether satisfactory results. Here again it is a matter of considering the way in which the vehicle will be driven: cornering techniques which feature numerous rapid modulations of the car's line, by the use of the accelerator pedal, cannot be employed satisfactorily or (sometimes) even safely with a turbocharged engine, but if the driver be content to employ point-and-squirt methods, the turbocharger may be quite satisfactory.

There is more to driveability than the specialised techniques of steering by throttle. There is the question of response in many other circumstances, whether in straightforward acceleration to overtake, in the readiness to supply full torque on demand after a few seconds on the overrun, or in the simple ability to idle at fairly low revs in a high gear, secure in the knowledge that immediate acceleration will be on tap if required, without a downward gearchange being necessary. In all these things the large lazy unsupercharged engine sets the standards by which the others may be judged: but where it is a matter of getting the desired performance from a given engine, then the mechanically supercharged version will be the best of the three possibilities, and the turbocharged version is likely to be the worst. Turbocharging has undoubtedly come a long way in the past few years, and it is now to be expected as a matter of course that any commercially available turbocharged engine will begin to give increased power as soon as the accelerator pedal is depressed — but unless it is already running at very high rpm, and has only been running against a closed throttle for a second or two, there will be a delay measured in seconds before the engine will be giving as much torque as it is capable of developing at the speed in question.

As we shall see in subsequent chapters, all these are essentially generalisations. It is possible to make a fairly highly tuned engine reasonably flexible; and it is certainly possible to supercharge an engine in such a way that it becomes very difficult to manage, the extreme being represented by erstwhile racing engines with centrifugal superchargers, such as the Miller and the BRM, both of which were plagued by an abysmal lack of power at low rpm, despite their phenomenal outputs at the other end of the speed range. Even more peculiar quirks may be identified in the turbocharged engine, for unless the turbine, its housing, the compressor and the engine all match perfectly, it is quite possible for the car to behave in a totally absurd fashion, perhaps going faster uphill or when towing a trailer than when running normally on the level, or perhaps accelerating pathetically until the engine reaches a higher rpm, whereupon the car leaps forward with such alacrity that the driver may find it hard to change up to the next gear fast enough.

Whether enjoying forced induction or not, the high-performance engine will also threaten its driver's equanimity with bouts of misfiring due to spark plug fouling, unless its ignition system is revised to cater for its special requirements. The high performance engine needs to have 'hard' or cool-running spark plugs if it is to run safely under full load; but such plugs have a tendency to foul and miss during and after protracted light running. The answer is some form of electronic ignition embodying the characteristics of capacitor discharge to give a very fast voltage rise at the plug points, the rise time (usually measured in milliseconds) being the time that it takes for the spark voltage to climb from 10% to 90% of its rated maximum. Capacitor discharge systems offer a rise time as little as two to four milliseconds, which is fast enough to spark and fire the charge in the combustion chamber, even though the nose of the plug be contaminated by deposits of oil or fuel. On the other hand it is usually a characteristic of capacitor discharge ignition that the spark duration is rather too short to ensure a satisfactory ignition of weak mixtures such as figure in part-throttle cruising; but there are one or two fairly complex systems available, notably the Mobelec, which combine a short rise time with a long spark duration. To be free of such complications, one must revert to the big lazy engine, which is invariably easier on its plugs and more often than not accepts a size that is relatively insensitive to heat — one of the many virtues of the run-of-the-mill American V8.

For road use there is yet another thing to be considered, and this is the matter of noisome emissions. The toxic emissions are usually measured as concentrations of carbon monoxide, oxides of nitrogen, and unburned hydrocarbons, in the exhaust gases: as a general rule, the highly tuned unblown engine will be the worst offender with oxides of nitrogen, the supercharged engine worst in unburned hydrocarbons, and the turbocharged engine generally the best all-rounder. A turbocharged engine is also likely to be the quietest, the emission of noise being no less important than the emission of undesirable chemical by-products of combustion. The process of expanding the exhaust gases through the turbine quietens the exhaust at the same time as it extracts power from it, that might otherwise go to waste. Undoubtedly the turbocharged engine is in many ways the most efficient — but, as we shall see in the next chapter, it all depends on what you mean by efficiency.

Chapter 2
Efficiency

The nonsensical rules that have governed most forms of motor sport, since the dim and distant years when its legislators might have been forgiven for not knowing any better, have unfortunately led most laymen to form the erroneous notion that the efficiency of an engine may be measured by comparing its power output with its displacement. Half a century ago it was already clear that this was an insupportable basis for comparison, with the Grand Prix cars created up to that time providing all necessary evidence. The 16 litre Fiat that won the French Grand Prix of 1907 developed 130 bhp; so did the 7.6 litre Peugeot that won in 1912, and so did the supercharged two-litre Fiat that won the Italian Grand Prix in 1924. Was any of these engines necessarily more efficient than the others? As the engines grew smaller, lighter, and more compact, they also grew more complicated, more difficult and expensive to make, and more intractable. More to the point, all three passed approximately the same mass flow of air in a given time, at full-power rpm which ranged from 1,600 in the biggest of them to 5,500 in the smallest. The point was made in the preceding chapter that power was directly related to air-swallowing capacity, and here was eloquent evidence: efficiency is surely a matter of what you get out of an engine compared with what you put into it.

Or is that only part of the argument? It is difficult to imagine the improvements in chassis design, in roadholding, braking, and so forth, exemplified by the later small cars being capable of realisation had the early gargantuan engines been retained to provide the necessary power. Whether for this reason or for other simpler ones, the fondness for high ratios of horsepower to engine capacity remains to this day, despite abundant further evidence of its irrelevance. The very highly supercharged V16 BRM 1½ litre engine produced as much as 585 bhp; the unsupercharged 3 litre Matra V12 eventually gave 500, and the 130 bhp that once involved four huge cylinders totalling over 16,000 cc, may nowadays be developed by a motorcycle engine with three little two-stroke cylinders mustering between them only 750 cc. If only specific fuel consumption figures could be obtained or deduced for all these engines we might have some very cogent things to say about their relative efficiencies, now that petrol and other fuels are not thoughtlessly to be wasted.

Power-to-displacement ratio may or may not be meaningless, and if efficiency be the object, who is to say that power-to-cost ratio should not matter, that power-to-fuel consumption should not, or that power in relation to space occupied by the engine might not, in some cases, be an equally important measure of efficiency, as it is in tanks, for example? As far as most cars are concerned, the necessary space can usually be found, and people are often prepared to pay at the outset for efficiency that can be enjoyed thereafter, while the actual displacement of the engine is or should be of even less concern than its external size, bearing in mind that two engines of similar proportions will vary in cylinder displacement in proportion to the cube of their respective lengths. In this connection it is surely significant that those two engines will vary in weight more nearly according to the square of their lengths, or if you prefer, to the square root of their displacements; and since the weight of a vehicle is the basic enemy of its performance, there is a very strong case for arguing that the true measure of efficiency should be the ratio of power to weight.

As between similar engines, the power-to-weight ratio is a fair criterion; but if two otherwise similar engines be differentiated by the provision of pressure charging for one of them, some very different results will emerge. The power-to-weight ratio of an unsupercharged engine is generally inferior to that of one given forced induction, and on this simple basis the pressure-charged engine is undeniably the more efficient. However, the greater heat flow through a blown engine is likely to demand a greater coolant capacity and a larger heat exchanger, all of which add to the weight of the total necessary installation, if not to that of the engine itself. Furthermore, the blown engine will have a higher specific fuel consumption, and must therefore be accompanied in the chassis by a greater quantity of fuel to last it a given time and distance; and this weight must also be brought into the calculations, though it is a weight that gradually disappears as the fuel is consumed and the products of combustion are unloaded into the atmosphere.

Already it is clear that there are two kinds of efficiencies that must be considered and set off against each other. One is volumetric efficiency, something in which the pressure-charged engine is clearly, and indeed by definition, superior to the normally aspirated one. The other is thermal efficiency, in which the pressure-charged engine, especially if the compressor be engine-driven rather than exhaust-driven, is equally invariably inferior to the normally aspirated one, and this relative inefficiency may cancel some or all of the gains made by the improvement in volumetric efficiency.

The first supercharged car in Grand Prix racing was this 1923 Fiat, which began with a troublesome Wittig vane-type compressor but was more effective with a Roots blower. In this form its engine deserves to be judged the most influential in the entire history of motor racing

For our purposes the volumetric efficiency of an engine is a measure of how well it breathes. At any given speed, the amount of air it takes in in one complete cycle may be more or less than the actual displacement of the pistons moving up and down their cylinders. Disregarding further considerations, the engine may be treated as a simple air pump, and its volumetric efficiency is simply a reflection of how much or little it is subjected to pumping losses. The artifices of the racing-engine tuner, who can contrive a volumetric efficiency better than 100% over a narrow speed range, have already been mentioned; now it is appropriate to list the various things that introduce pumping losses so as to reduce the volumetric efficiency of the average engine throughout its working range, or the racing engine outside its carefully tuned power band, to a level likely to be lower than 80% and perhaps as low as 60%.

19

I here are such things as restrictions in the inlet manifold, the size and shape of the ports, restrictions within the carburettor, poor exhaust scavenging (often caused by excessive back pressure which prevents the cylinders from being evacuated during the exhaust phase), inlet valve design,and of course the camshaft which controls the opening and closing of the valves, and thus determines the duration of the period in which the other restrictors can be effective. Every little constriction, every aerodynamic step over which the incoming air may trip and burble, is an impediment to the flow of air that can only be induced by the existence of a pressure difference between the inside and the outside of the engine: and however effectively the cylinders be evacuated during the exhaust phase, so as to lower the absolute pressure within them, the unsupercharged engine is always limited by the natural maximum external air pressure which is that of the ambient atmosphere.

However, if the inlet manifold is pressurised by supplying it with compressed air from another pump (the supercharger) all these impediments may be overcome: instead of a pressure difference between the inside and outside measuring somewhat less than 1 atmosphere, it may be made substantially higher. All other things being equal, the greater the pressure difference, the greater will be the mass flow into the cylinder. The ratio of the mass of new charge in the cylinder to the mass of charge that would suffice to fill the displacement volume at ambient atmospheric pressure, is a measure of the volumetric efficiency of the engine and its complete respiratory system. Volumetric efficiency of the engine is generally expressed as a percentage, and in a naturally aspirated one it seldom reaches a figure higher than 90%. In the case of an engine highly tuned to exploit resonance and ram effects, the figure may rise at certain operating speeds and wide throttle openings to higher than 100%, perhaps even 110%, and very occasionally more; but a pressure-charged engine can be brought up to figures very much higher, and in fact achieves 100% volumetric efficiency in the part-throttle cruise condition with the manifold pressure (or boost pressure, as it is often called) only slightly above zero.

If the strength of a supercharged engine is in its volumetric efficiency, its weakness is in thermal efficiency. It is worth making clear at the outset that no internal combustion engine is very good in this respect, and as a general rule external combustion engines are very much worse. Only in exceptional cases is this situation reversed: with vast and elaborate heat exchanger facilities built at enormous expense and considerable pains, the thermal efficiency of a steam turbine electricity generating station may in a few modern examples reach an efficiency as high as 45%, but such cases are rare. Among automotive engines, the diesel often appears attractive despite its numerous serious shortcomings,by achieving a thermal efficiency of about 40%, while the naturally aspirated petrol engine seldom does much better than 33%. In other words, of all the heat units represented by the calorific content of the fuel ingested by the engine, only one third are converted into useful work. Of the remainder, half is voided to the atmosphere through the exhaust system, and the other half is absorbed by the engine, raising the temperature of its components which then shed the heat as best they can, until it, too, is eventually wasted by the process of radiation and transfer to the surrounding and cooling air.

Alas that we shall never see such naturalistic motor racing again as here at Pescara in 1938. The Mercédès-Benz V12 gave 468 bhp from 3-litres at 2.2 atmospheres boost

When an engine is supercharged, a greater quantity of heat units flow into it, in direct proportion to the increase in the system's volumetric efficiency. When the calorific content of the fuel in this intensified charge is liberated as heat during combustion, it is only to be expected that a similarly greater quantity of heat will be absorbed by the engine; and it is a readily observed fact that component temperatures reach much higher levels in a supercharged engine than in a naturally aspirated one. It is seldom that this problem can be solved merely by increasing the overall capacity of the cooling system: all this does is to maintain the coolant temperature at the same level, but the paths that the heat takes from this source of generation to the coolant are not thereby made any easier. Consider, for example, the heat fed into the piston crown during combustion: some of it escapes by radiation from the underside of the piston, some is conducted to oil mist or spray making contact underneath the piston, and a little is transferred to the next fresh intake charge; but the bulk of it has to travel through the metal of the piston crown to the ring belt, whence it passes through the piston rings to the cylinder walls, and thence to the surrounding coolant. Clearly, any increase in the amount of heat fed into the piston crown will cause a steepening of the temperature gradient between that input area and those areas through which the heat may be allowed to escape; and if these areas are not increased, they will simply have to survive operating at higher temperatures.

It is not only piston rings that are affected, but numerous other components, notably valves, valve seats, and spark plug points, as well as any relatively sharp edges in the combustion chamber formed by squish bands or valve clearance pockets in the piston crown. If any of these small but critical areas becomes overheated, it then provokes pre-ignition of the engine which in turn leads to a catastrophic pressure rise, similar to detonation, which can wreck an engine in a matter of seconds.

The problem of detonation must also be approached from another angle intimately related to thermal efficiency. The internal combustion engine is simply a heat engine, one in which heat is converted into work. The fresh charge in each cylinder is compressed by the rising piston, and the work thereby done on it is converted into heat, raising the temperature of the charge to the point where combustion initiated by the ignition system proceeds as efficiently as possible. During this combustion, the calorific content of the fuel is liberated as heat, and the process causes the pressure of the gases within the combustion chamber to rise rapidly, forcing the piston downwards again. This is a mechanistic way of viewing the conversion of heat into work; the thermodynamicist looks at it slightly differently, reasoning that as the piston descends, during the working stroke, it causes the volume of the cylinder to increase, and by expanding the heated gases therein, they are cooled, the heat loss being converted into useful work. The design of the internal combustion engine is such that there are practical limits to the degree of expansion that can thus take place: in a typical engine with symmetrical valve timing, the expansion ratio is equivalent to the compression ratio. There have been engines made to work on slightly different timings that give a greater expansion ratio than compression ratio, but in practice these have been found suitable only for constant-speed operation in large stationary installations. In fact the ordinary normally aspirated car engine is in some slight measure akin to these hyperexpansion engines, because, although its valve timing is likely to be more or less symmetrical (that is, the exhaust valve opens as many degrees before bottom dead centre as the inlet valve closes after it, the volumetric efficiency of such an engine is generally less than 100%, which means that the true ratio of compression of the charge is proportionately less than the geometrical compression ratio, as determined by the volume of the combustion space, and of the cylinder.

As we have seen, the volumetric efficiency of the supercharged engine may be considerably higher than 100%. In other words, the actual compression of the intake charge is considerably greater than the geometrical compression ratio, determined by the cylinder and head dimensions. From the thermodynamic point of view this is all to the good — but the expansion ratio remains unchanged, and so there is no possibility of recovering during the expansion phase as much energy potential as was created during the increased compression phase. It is because of this simple imbalance between the overall compression ratio and the expansion ratio that the thermal efficiency of the supercharged engine is so much lower than that of the naturally aspirated one.

It follows that much of the wasted heat escapes down the exhaust pipe when the exhaust valve opens. Instead of being wasted it could be harnessed by further expansion, through

The 1939 Mercédès-Benz, first to exploit two-stage supercharging, developed 483 bhp from 3-litres at 2.6 atmospheres boost

some device that could convert the remaining heat, or some of it, into useful work. This can be done in an exhaust turbine.

There are several ways in which the work extracted at the exhaust turbine can be harnessed. The turbine shaft could, for example, be connected by suitable gearing to the engine's crankshaft or main output shaft, but this raises further difficulties, some of which are of a mechanical nature, others thermodynamic. In the first place the exhaust turbine is by nature a component that rotates at a very high rate indeed, much higher than that of any engine crankshaft, and if the two are positively geared together there will be severe inertia loading problems when the engine is accelerated or decelerated. The most satisfactory answer to this problem is to interpolate a fluid coupling, and this has been done in some engines that will be examined in Chapter 8.

The other aspect of the problem is the way in which the exhaust turbine responds to input. It is responsive to the mass flow of exhaust gases through it, and this increases in proportion to the engine load. When the engine is operating at full load with wide-open throttle, its volumetric efficiency approaches its maximum, and so correspondingly does the effective overall compression ratio — from which it follows that the maximum opportunity is created for exploitation of further expansion subsequent to that which takes place with the cylinders. When the engine is running against a light load, however, it is throttled down and the volumetric efficiency is decreased accordingly, as is the mass flow through the system: so at low speeds and small throttle openings, there would be very little work done on the turbine, perhaps less than is necessary to match the mechanical losses due to friction in the bearings and gearing or other mechanical connection between the turbine and the engine output shaft. In other words, the engine would then have to drive the turbine, with a resultant net reduction in thermal efficiency.

It might therefore seem more attractive to dispense with such a coupling arrangement and instead to use the turbine for driving such things as the coolant water pump (or cooling air fan in the case of an air-cooled engine) since the demands on the cooling system are proportional to the engine load in the same way as is the response of the exhaust turbine. In fact the power absorbed by the coolant pump or fan is so small that the bother would not be justified. Where the exhaust turbine comes in useful is in a pressure-charged engine, when it can be coupled directly to the compressor. Thus we create the turbocharged engine, and from what has already been said, it must be clear that this is thermally much more efficient than the engine with mechanically driven supercharger. The engine's requirements from the compressor are proportional to its load, and so is the output available from the exhaust turbine. Instead of taking power from the engine crankshaft to drive the compressor, the power is abstracted from the exhaust gases where it would otherwise be allowed to go to waste, and the net saving is realised as extra power available from the engine output shaft. As the load and the volumetric efficiency and the effective compression ratio are increased, so is the effective expansion ratio: so the turbocharged engine can achieve levels of thermal efficiency considerably higher than the mechanically supercharged type, approaching comparison with the thermal efficiency of a naturally aspirated engine.

If it were possible to make such systems entirely efficient, then the turbocharged engine would be thermally as good as the unblown engine, but nothing is ever as perfect as that. In any case there are other factors limiting the designer's freedom to exploit ever higher overall compression ratios; and, as we shall see, these are linked to the inefficiencies that we have just been lamenting.

The most critical factor is the propensity of the fresh charge to detonation. As we have already seen, this can be initiated by pre-ignition as a result of some localised hot spot within the combustion chamber: while the charge is still undergoing the normal mechanical compression, as the piston approaches the top of its stroke, pre-ignition of a

Simple Roots supercharger on the nose of the M125 straight-eight which gave the 1937 GP Mercedes-Benz as much as 646bhp — and about 140 at a mere 1500 rpm

portion of the charge causes a flame front to expand from the point of ignition, ahead of a rapidly expanding mass of burning gases, and these simply compress the remaining charge into an even smaller space, until the stage is reached where the charge is compressed and heated to the point where it detonates spontaneously. Pre-ignition is not a necessary prelude to detonation; excessive compression and heating is all that is required. It cannot be too strongly emphasised that with normal engines and fuels it is detonation which sets the limit to performance: where an unsupercharged engine is tuned for maximum output, the compression ratio is adjusted so that the combustion chamber pressure at the instant of ignition is as high as possible without running the risk of detonation occurring. Just what that pressure is depends on a number of design features including the shape of the combustion space, the materials by which it is bounded, and the presence of hot spots — including incandescent deposits formed on its surfaces during its working life; but the most critical factor is the fuel itself, which may vary considerably in its anti-knock value or octane rating.

25

Turbocharging and Supercharging for Maximum Power and Torque

When the engine is supercharged, most of these considerations remain unaltered, and so therefore does the maximum pressure that can safely be allowed to build up in the combustion chamber. In the first chapter we saw that the mean effective pressure of a supercharged engine may be substantially higher than that of an unsupercharged engine, but now we see that the peak pressure can be no higher. The difference in mean pressure must therefore be due to the fact that the rate of pressure drop during the expansion phase is lower — as is confirmed by the fact that there remains so much scope for further expansion of the combustion gases after the exhaust valve has opened. This explains one of the less appreciated virtues of the supercharged engine, the fact that the mechanical loading on the engine's components does not increase at the same rate as the power. The limits to the mechanical integrity of the engine are imposed by inertia loadings and peak combustion pressures, and these are not significantly different in supercharged or unsupercharged engines. In fact the stress reversals in a supercharged engine may be less severe, not only because of the more sustained pressure on the piston during the expansion stroke, but also because the positive inlet manifold pressure of the supercharged engine causes a higher pressure to be maintained in the cylinder, and therefore on the piston crown, during the induction stroke. Indeed, the only reason why inertia loads might be higher in a supercharged engine is that it might be necessary to increase the mass of the piston in order to improve the flow of heat through it, and to prevent the temperature gradients within it from becoming too steep.

Coming back to the problem of limiting the compression pressure to avoid detonation, we can see that if the limit is reached in an unsupercharged engine by raising the geometrical compression ratio as high as is safe, then in a supercharged engine of so much higher volumetric efficiency it will be necessary for this geometrical compression ratio to be reduced. This disappointment is softened only by the fact that it makes it easier to achieve a satisfactory shape in design for the combustion space, which should ideally have the least possible surface area in relation to its volume. In an engine with a high compression ratio this is very difficult to achieve, because of the conflicting requirements of adequate valve diameter and lift, with the result that the combustion space usually ends up as an irregular shape of fairly large diameter and relatively little depth, with a consequently high ratio of surface to volume. The lower piston crown demanded by abasement of compression ratio, necessary in a supercharged engine, increases the depth of the combustion space and thus improves its shape and surface-to-volume ratio, which in turn permits more efficient and better controlled combustion, with less risk of dangerously rapid pressure rise during the critical period when the danger of detonation is at its greatest. On the other hand, that same risk is made more serious by those shortcomings in efficiency to which we referred a few pages earlier; and this brings us to a consideration of a third type of efficiency that is crucial to the whole business of supercharging — adiabatic efficiency.

Efficiency is always a measure of something related to something else, and in this case adiabatic efficiency is the ratio of adiabatic work to actual work — or if you like, the ratio of the ideal to the practical. The word *adiabatic* may be unfamiliar, and is best

2,000 bhp is claimed by owner Dennis Priddle for this 'Funny car' engine. This Donovan Chrysler V8 of around 8-litres using a Roots type blower taken from a GMC truck engine, is designed to produce excess power for very short periods of time and is the standard wear in drag racing

explained by defining an adiabatic operation as one in which no gain or loss of heat to or from external sources occurs. It follows that all work done must be done at the expense of the internal energy, and is equal to the difference between its initial and final values — or to put it another way, that if a compression or expansion is to be adiabatic, the operation must be reversible, so that the initial state can be restored by reversing all the changes.

Alas, this ideal cannot be reached, for it is impossible to construct an engine working in a complete cycle which does work and receives heat only: in other words it is impossible to transform all the heat added to an engine into work, which is why our engines have such a depressingly low thermal efficiency. No process is reversible that is accompanied by friction, by an unrestricted expansion, or by a heat transfer resulting from a temperature difference; and all these phenomena are inseparable from our engines.

Turbocharging and Supercharging for Maximum Power and Torque

Let us see how this affects the volume of gas being compressed, as in the case of the air or mixture delivered by a supercharger, or (however it arrived there) the fresh charge inside an engine cylinder at the end of the induction phase. When this volume of gas is compressed, and cannot escape, it is obvious that the gas pressure rises in proportion to the degree of compression; but we also know from practical experience as well as from theory that the temperature of the gas rises too — but it does not in practice rise in strict proportion. The reason for the difference is that whereas the gas is physically trapped, heat is not. If no heat were to enter or escape during the process of compression, then the compression would be truly adiabatic; but in fact heat can be given up to or taken from the walls of the vessel or duct in which the gas is compressed, and in practice the result is that the charge finally compressed in the cylinder reaches an appreciably higher temperature than it should in a truly adiabatic process. The measure of efficiency of a compressor is how nearly it comes to the true adiabatic state: so let us consider the case of a compressor of 50% adiabatic efficiency. The final temperature of the compressed gas will be twice that of a true adiabatic compression, the extra heat corresponding to losses due to the low pumping efficiency of the compressor.

This has a deleterious effect on the engine's power output, which depends on the mass of charge that can be consumed in a given time. At any given pressure the density of a gas is inversely proportional to its temperature: and so the rate of charge is reduced by the heating to which it is subjected. When the charge finally reaches its point of ignition in the combustion chamber, it is the *weight* of the charge that will determine how much essential energy can be extracted there and converted into useful work — but it is the *temperature* of the charge that determines the onset of detonation.

This temperature may be calculated with the aid of the formula

$$T_2 = T_1 \left(\frac{P_2}{P_1}\right)^{0.283}$$

In this expression the temperatures and pressures are absolute: thus T_1 is the inlet temperature in degrees R, T_2 is the output temperature on the same scale, P_1 is the inlet pressure (normally one atmosphere) and P_2 the discharge pressure from the compressor. However, this formula applies to a compressor working ideally with 100% adiabatic efficiency; this ideal being unattainable in practice, the actual temperature rise must be calculated by dividing the difference between T_2 and T_1 from the above equation, by the adiabatic efficiency of the compressor.

To postulate an example where a compressor of 50% adiabatic efficiency takes in air at an ambient temperature of 20°C and delivers it at a pressure of 1.5 atmospheres, the delivery temperature will be 90.0°C — whereas if the compressor were of 75% adiabatic efficiency, its outlet temperature would be only 67.3°C.

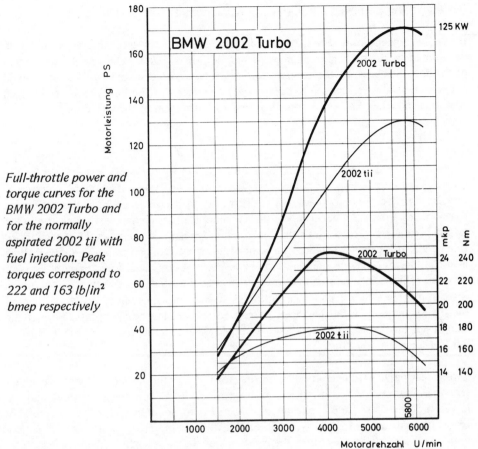

Full-throttle power and torque curves for the BMW 2002 Turbo and for the normally aspirated 2002 tii with fuel injection. Peak torques correspond to 222 and 163 lb/in² bmep respectively

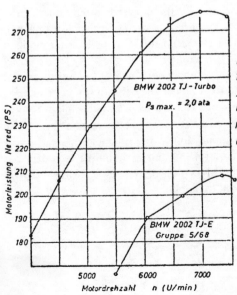

For racing, the BMW 2002 in unsupercharged form (as the TJ–E) could be made to give 208 bhp, more than the road going 2002 Turbo; but the racing TJ–Turbo matched this at a mere 4,500 rpm, and went on to give an additional 70 bhp

29

Turbocharging and Supercharging for Maximum Power and Torque

It can be argued from this that the adiabatic efficiency of the compressor offers, in conjunction with the degree of boost it supplies, the necessary clues to the calculation of the amount by which the compression ratio of an engine should be reduced to avoid detonation — always supposing that for the fuel concerned the limiting unsupercharged compression ratio is known. A popular formula for determining this reduction is

$$R_2 = R_1 \left(\frac{P_1}{P_2}\right)^{0.5}$$

where R_1 is the unsupercharged compression ratio and R_2 the compression ratio for the supercharged engine. Popular though this formula may be, it should be used with caution, for it is unsatisfactory in some ways. In the first place it takes no account of the differences in volumetric efficiency between the supercharged and unsupercharged states, and in the second place the power to which the inlet manifold pressure ratio is raised was established empirically in the days when 50% adiabatic efficiency was the norm for compressors then in use. It seems desirable that the expression should be adjusted to suit compressors of different efficiencies: the frequent occurrence of overheating or destructive detonation in engines supercharged in accordance with the conventions of this ruling may be accounted for by the failure of the engineers concerned to take the pumping efficiency of the compressor into consideration, when determining the desirable reduction of the engine's internal compression ratio.

Before leaving this subject it is interesting to examine the device of two-stage supercharging, whereby the adiabatic efficiency of the total compressor system may be increased. This technique is applicable to blowers having no internal compression (notably the Roots and the centrifugal types, to be described in the next chapter), which simply displace air and deliver it against the resistance of the air already in the inlet manifold. It follows that the work done, or power absorbed, by the blower is the product of the volume it displaces in a given time, and the contra pressure against it which it operates. However, if an intermediary blower of smaller displacement (a suitable relationship is that the displacement of the second stage blower should be about 70% of the first stage) be inserted in the system, then each of the compressors will be working against significantly smaller contra pressures, and the total power they absorb will be reduced — typically by more than 20%, which in the case of a Roots blower would be equivalent to raising the pumping efficiency of the system to a level competitive with a single-stage centrifugal blower. The first really effective application of the principle to a car engine was by Daimler-Benz in their Grand Prix cars of 1939, but it should be recorded that Chadwick applied the principle to his car in 1909 with a three-stage centrifugal blower exhibiting crude construction but very advanced thinking.

30

Chapter 3
Compressors

The compressors that are or have been used in conjunction with automotive engines may be divided into two classes, those which give a positive displacement of a known volume of air in each complete cycle, and those that do not. These two classes are susceptible of further subdivision: the first may be divided into those that work by internal compression and those that do not, while the second class can be divided into those based on radial flow of air, and those based on axial flow. Each type has its own particular advantages and disadvantages, which may suit it or disqualify it for a particular kind of duty. For that matter, each type may be seen to have been used at some period in history simply because it was then fashionable, regardless of its suitability for the task.

The number of different kinds of positive-displacement blowers is legion, for it must be understood that any machine capable of functioning as a heat engine may equally well, after deletion of the facilities for introducing fuel and igniting it, serve as a pump. Thus some of the earliest superchargers tried in racing cars were of the simple reciprocating piston type, to be found most notably in the early experiments by Birkigt of Hispano-Suiza, and more surprisingly in certain racing two-stroke motorcycles of the 1930s. Despite its effectiveness as a compressor, the piston type of blower is wholly unsatisfactory for duty in a high-performance car: its bulk represents a considerable nuisance, its mechanical friction an unnecessary burden, and most of all the necessary sheer size of the pistons and the motions of the connecting rods by which they are coupled to the crankshaft, impose limitations (not only of inertia loading but also of rubbing speed, as represented by piston velocity) that impose a severe handicap on the engine's freedom to run to those respectably high rates of revolution where power is to be sought.

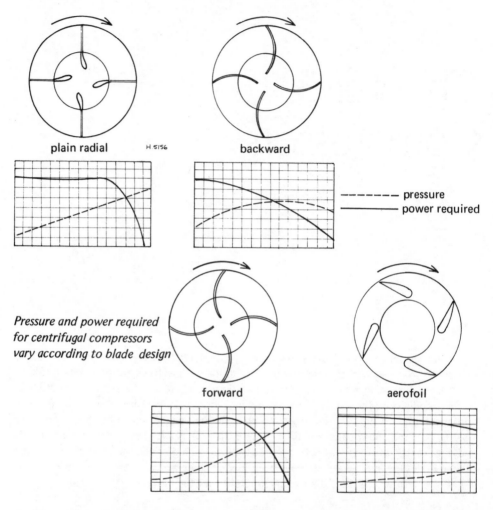

plain radial   H 5156   backward

- - - - - - - pressure
——————— power required

Pressure and power required
for centrifugal compressors
vary according to blade design

forward   aerofoil

A nice modern example of the freedom with which positive displacement machines can be treated as engines or as compressors is afforded by the Wankel engine. Today this is viewed simply as a power unit; but before it was developed as such, it served most effectively and efficiently as a supercharger for a number of very small but astonishingly powerful motorcycle engines used in a record-breaking spree by NSU. Those days were before the clever engineer Dr. Fröde (or Froede) of NSU performed his kinematic inversion of the Wankel rotary machine, to create the essentially simple configuration by which we know it today. At that time both the rotor and the outer casing revolved, and by accepting very high speeds of rotation and consequently high rubbing velocities at the tip seals of the rotor, NSU were able to achieve very high boost pressures combined with respectable adiabatic efficiency. It would be possible to employ the modern configuration of the Wankel machine as a compressor and, by accepting the short life of the machine when run at very high speeds, to achieve comparable pressures and

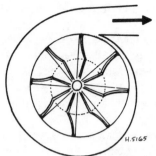

Centrifugal compressor-flow accelerator (examples: Miller, Duesenberg, Lycoming, BRM, Holset, Eberspacher and Rajay)

Axial-flow compressor-flow accelerator (example: Napier)

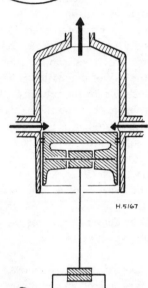

Roots compressor — positive displacement: no internal compression (examples: GM, Wade and Marshall)

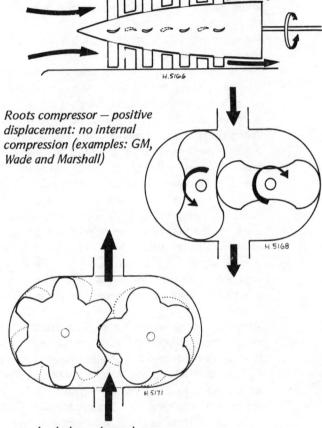

Piston compressor — positive displacement: internal compression (examples: Hispano-Suiza and DKW)

Lysholm or 'screw' compressor — positive displacement: some internal compression (examples: BroomWade and Elliott-Lysholm)

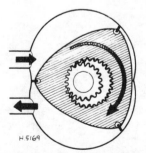

Eccentric vane — positive displacement: internal compression (examples: Shorrock, Zoller, Cozette, Powerplus and Centric)

Wankel compressor — positive displacement: internal compression (example: NSU)

efficiencies; but for practical long-term use it would be essential (pending the development of less troublesome seals than have at present to be accepted) to limit the cyclic speed of the Wankel — and even then the need to lubricate the seals would introduce the risk of oil being entrained with the charge and introduced to the combustion chamber, where it would have a dangerously degrading effect on the resistance of the fuel to detonation.

The same objections apply to nearly all displacement blowers that work by internal compression. Most of these machines are now outmoded: the Centric, the Zoller, the Cozette and their like have all been rejected after extensive trials in the 1930s, because they all incorporated some out-of-balance mechanism that prevented them from being run at high cyclic speed, and therefore made it necessary for them to be large and bulky to compensate for this disability. Furthermore, all these machines involved rubbing and sliding contacts that demanded lubrication, creating the risk of detonation at high speed already described. Even in compression-ignition engines, where the presence of the oil does not create such hazards, it is still a source of problems, because the oxidation and general breakdown of the oil on exposure to high-temperature air produces undesirable deposits on valve stems, heads, and other internal surfaces of the engine. The same objections apply to the only eccentric-vane type of compressor now surviving, the one known as the Shorrock. The only advantages to be claimed for it are a somewhat higher adiabatic efficiency than the Roots type of blower, the machine being able to reach 55 or 60%, while it is also more suitable to the delivery of high boost pressure than a single-stage Roots blower, which does not work very well at delivery pressures higher than about 1.6 atmospheres. In every other respect the Roots type is preferable.

Although coming within the positive displacement category, the Roots blower has no internal compression. It works rather like a gear-type oil pump, with two contra-rotating elements that alternately occupy and evacuate a prescribed volume. Connected solely by their drive gears, these two rotors do not actually touch each other, though they are machined to intermesh with great accuracy. Accordingly there is negligible friction, and internal lubrication is quite unnecessary. Furthermore, the rotors are balanced about their axes of rotation, and thus can be driven at very high speeds, the limit being reached when the air can no longer pass through the intake port at a higher rate of mass flow.

However close the clearances between the rotors and their casing, some leakage from the delivery to the suction side is bound to occur, and this results in pumping losses that assume greater significance as the speed of rotation is reduced. The delivery characteristics of the Roots blower are therefore such that the boost pressure increases with engine speed up to a certain level, beyond which the increase becomes less notable, and eventually the boost pressure will be substantially constant in the topmost portion of the working speed range.

There are some disadvantages to the type, certain of them being of concern only to the manufacturer. For example, the inlet side is kept cool by the flow of cold air, while the

Chart showing relationship between density and pressure ratio with centrifugal compressor

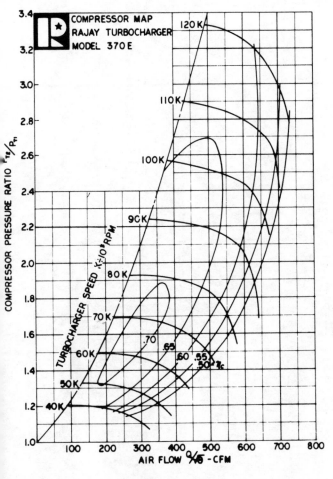

Typical centrifugal compressor flow map

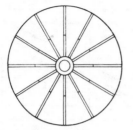
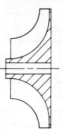

Simple compressor impeller with straight blades

Compressor impeller with curved inducer

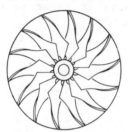
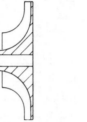
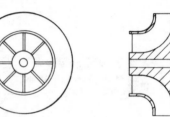

Compressor impeller with backward curved blades

Shrouded impeller

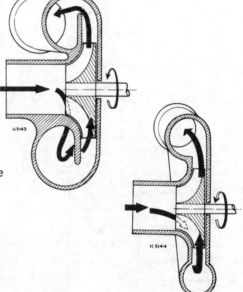

Compressor with parallel-wall diffuser and toroidal collector

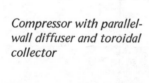
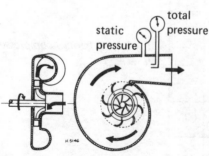

total pressure

static pressure

Compressor with vaned diffuser

Compressor with parallel-wall diffuser and scroll or involute housing

delivery side is hot because of the rise of temperature through the machine, whence it follows that if the casing is bored circular when cold, it will distort in operation, and bad leakage will ensue. This would be particularly serious because the leak would be of high-temperature air back to the suction side, where it would increase the inlet temperature, and hence the delivery temperature, which again leaks back at a still higher temperature to create a vicious circle. In practice the casing is machined eccentric so that in operation it will be suitably symmetrical.

It used to be an objection against the Roots blower that its delivery was of a pulsating nature. Where the two rotors each possessed two lobes, there would be four impulses for revolution, and so a steady flow could not be secured unless the volume of the manifold-ing between the blower and the cylinder head were great enough to damp out these pressure oscillations. Some Roots blowers have rotors with three or four lobes, creating a high-frequency pulsation that may be easier to deaden; but the most effective solution of the problem is either to machine the rotor lobes in a helical form (which is expensive) or to shape the ports after the fashion of a helix, a cheap way of achieving the same effect.

The adiabatic efficiency of a Roots blower seldom exceeds 50%, but we have shown in the preceding chapter how this can be effectively increased by interpolation of a second blower. The same two-stage technique can be employed to secure effective operation at a higher boost pressure than the 1.6 atm limit generally recognised as applying to the type. For higher pressures from a single blower, positive displacement has to be combined with internal compression, on the lines of the eccentric-vane machines already mentioned, in which air is trapped between two more or less radial blades or vanes and compressed, as it is carried by them around the casing, into a smaller volume. There is, however, a blower that has much in common with the Roots type mechanically, while its performance characteristics are those of an eccentric-vane type: this is an extremely expensive supercharger known as the Elliot-Lysholm, often used for supercharging large stationary or marine engines but that scarcely ever can be judged worthy of its cost for use in cars. Briefly, its two intermeshing rotors are of disparate size and shape, the larger one having 'male' or convex lobes, and the smaller having 'female' or concave ones. The lobes are machined in helical form, and as the rotors turn, they draw air lengthwise through the casing, and progressively compress it before discharging it at a point diagonally opposite its entry. All the advantages of cleanliness, lack of friction, and inherently perfect balance, are retained from the conventional Roots blower, together with a suitability for working at higher delivery pressures and higher adiabatic efficiency such as would be respectable even in a vane-type blower. Indeed, large Lysholm compressors have displayed an adiabatic efficiency as high as 80%.

Supercharger efficiency of this order is generally associated with blowers that are not of the positive displacement type, but are simple flow accelerators. The best known in this application is the centrifugal compressor, such as has been almost universal in the supercharging of piston aero-engines, has been employed in engine-driven form for sports and racing cars made by Miller, Duesenberg, Auburn and BRM, and figures in all automotive turbochargers. Needless to say, the aircraft industry has developed this type of supercharger to an extraordinarily high level of efficiency, following the lead given by an RAE scientist named James Ellor, who subsequently did valuable work for Rolls-Royce in the 1930s and after. Inevitably some of the refinements of the aero-engine supercharger are beyond the purse if not the perception of automobile engineers, but even the simpler types of centrifugal compressors used in cars are notably efficient compared with their rivals.

In construction, the centrifugal compressor is very simple, consisting of an impeller driven at a high rotational speed so as to rotate inside a compact casing. The air enters the casing and approaches the impeller along the axis of the drive shaft, and leaves the impeller radially at its circumference, travelling with high velocity imparted by the vanes projecting from the face of the impeller. The air leaving the impeller passes into a diffuser where its velocity is reduced through the reaction of a gradually increasing cross-sectional area of the passage, so that the kinetic energy of the air is converted into pressure energy, the air being then ducted away from the diffuser at the same mass flow rate as it entered the impeller, but at substantially higher pressure and temperature.

Impeller design varies quite a lot. The cheapest and easiest to make are those impellers that have perfectly straight radial blades, but better efficiency is achieved if the leading edge of the blade at the inlet nozzle is curved to induce a cleaner flow of air into the impeller. This detail improvement, which considerably increases the difficulties of manufacture, improves efficiency by reducing the shock losses at the inlet. Thereafter the blading may be straight and radial or curved backwards, or forwards, and each has its advantages. The magnitude of the pressure ratios produced by the three impellers is in the order radial blade highest, and forward-curved blade lowest: while the velocity at the impeller exit is highest with the forward-curved blade and lowest with the backward-curved. Actually, if the velocities at the impeller exit could be converted completely into pressure, the overall pressure rise would be highest with the forward-curved blade, and lowest with the backward-curved; but the process of converting kinetic energy into pressure energy is not generally an efficient one. Therefore, if the backward-curved vane imparts the least kinetic energy to the air, the losses in the diffuser would be least (supposing equal efficiency of the diffuser in all cases) and they would be greatest for forward-curved blades. In practice, the radial-blade impeller is generally preferred because it has a higher pressure ratio to begin with, and because it is very much easier to make — the speeds of rotation are so high that the tremendous centrifugal forces generated tend to load the extremities of curved blades as though to straighten them, and this may lead to failure in the metal. A radial blade is not susceptible to this kind of distortion.

Where the cost of making a backward-curved set of blades can be accepted, this type of impeller rotor is the most efficient, the forward-curved blade being still the least attractive. The small deficiency of backward-curved blades as regards pressure ratio can easily be made up by a slight increase in speed of rotation.

How slight this increase need be, is revealed by a study of the output characteristics of the centrifugal blower, which are extremely speed-sensitive. The output pressure in fact increases as the square of the rate of rotation, which means that a supercharger delivering 5 atm boost at 50,000 rpm will give only 1.25 atm at 25,000. The effect of this upon engine output and upon the sheer manageability of the car's performance can be critical to the whole operation, as it was in the case of the V16 BRM in the early 1950s. In that car, raising the engine speed from 7,000 to 9,000 rpm increased the power output by no less than 80%, so if the driver sought to initiate a drift by the methods appropriate to the handling characteristics of Grand Prix cars of the period, opening the throttle actually in a corner, he could find himself within the wheelspin range at one moment and way beyond it at the next, even though the car might have gained only a little in speed during that brief interval — and once the wheels started to spin, instead of having a tendency to regain their grip as the engine speed continued to rise (as would happen with the back-up torque curve produced by more conventional engines) they would proceed to spin even more wildly. The two-stage centrifugal supercharger designed and made by Rolls-Royce for the BRM attracted far more contumely than it deserved: it made the engine extraordinarily powerful, its output in the final stages of its development amounting to no less than 585 bhp, this figure being determined by the airflow limits imposed by available carburettors. In this last form, the supercharger delivered air at 5.7 atm and a rate in excess of 1 lb per second, a performance that could not have been approached by any other available compressor, even had space been available. The BRM supercharger was extremely compact, measuring less than five inches in length and twelve in casing diameter, and at the same time it achieved very high adiabatic efficiencies, an important consideration when a 1% variation in efficiency can affect engine output by 5%.

In fact the difficulties of controlling the BRM would have been greatly eased had the method proposed by Rolls-Royce for controlling the supercharger been adopted. This was based on the principle of vortex throttling, which will be described in Chapter 6. The idea was to vary the output of the supercharger so as to maintain a sensibly constant boost pressure over a satisfactorily wide working range of engine crankshaft speeds. In the absence of the vortex throttling device, the BRM was plagued by the boost pressure's acute sensitivity to speed.

There has of course to be a limit somewhere to the continuing increase in output of this type of blower as it is accelerated. The generally accepted limitation is on the permissible blade tip velocity, for the problems of transonic air speed (made familiar to us all by those adventurers in aviation technology who first sought to fly at the speed of sound) begin to intrude as the tip speed approaches 1,000 ft/sec. It used to be thought, pursuing the same reasoning as was once applied to air speeds, that the speed of sound (about 1,070 ft/sec) could not or should not be exceeded at the blade tips; more recently ways have been found to permit this, and some impellers have been run at peripheral speeds of about 1,300 ft/sec. However, the losses suffered by the machine may then be very serious, shock waves playing havoc with the air flow, which at sonic speed no longer has the characteristics of a compressible fluid such as a gas, but begins to act like an

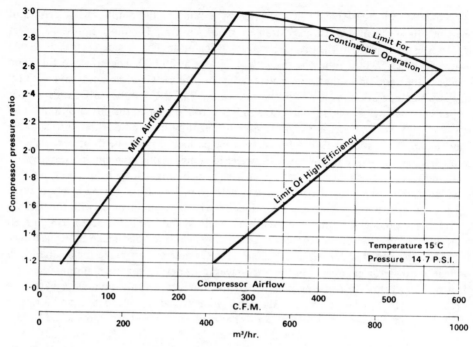

Air flow range of the modern 3 inch Holset 3LD turbocharger

incompressible one such as water, with the result that the conversion of kinetic to pressure energy in the diffuser section is greatly upset.

There is another way in which the airflow through a centrifugal supercharger may resemble that of water in a centrifugal pump, and this is when surge occurs. The problem is one that is peculiar to the class of compressors that do not feature positive displacement, and is a violent and often audible air disturbance that can reach such magnitude that the machine may be severely damaged. The surge occurs when the air intake to the compressor is throttled down below some critical point that varies according to the mass flow and pressure ratio. Any further attempt to reduce the flow through the impeller results in an instantaneous reversal of flow, the air travelling up the face of one diffuser vane, turning away in a turbulent manner and coming back along the opposite face of the adjacent vane. The condition is extremely unstable and may amount to anything from a faint pulsation, scarcely detectable, to a violent juddering that can destroy not only the supercharger but also the engine. The rate of heat build-up in the compressor is so rapid that the impeller may melt, while such air as is delivered to the engine will likewise be dangerously hot and will almost certainly cause destructive detonation within the cylinders. Alternatively the air may not get through at all, but be completely stalled, where conditions can become such that all the power supplied to the impeller is used up in internal losses.

If a series of curves is plotted for various impeller speeds, relating delivery pressure against mass flow, each curve will be found to end abruptly at some definite measure of flow at which surge begins. These points will be seen to lie approximately along a curve originating with the cördinate axes, and the entire area to the left of this curve represents a region in which the compressor cannot be stably operated. By changing the diffuser design it is possible to vary the characteristics over a fairly wide range for a given impeller, the phenomenon being identified as a function of the impeller speed and the diffuser shape and area. This reasoning led to the development of compressor housings incorporating vaneless diffusers. Compressors of this type cannot attain the very highest efficiencies, but are indeed often deliberately designed to sacrifice a little of the peak efficiency in exchange for a broader operating range. The diffuser then becomes either a pair of parallel walls leading into a cavity or (as is generally preferred because it is smaller and lighter) a housing in the shape of a scroll or involute. Compressors of these kinds are very much less sensitive to surge than those with diffuser vanes, and commonly have an extremely broad flow range. They are usually characterised also by a low peak efficiency, but some of this loss can be restored by backward-curved blading on the impeller. Clever design along these lines may limit the sacrifice in peak adiabatic efficiency to as little as 3%.

Even more beset by surge sensitivity than the centrifugal compressor is the axial-flow type, in which a series of rows of radial blades of airfoil section propel the air along an approximately cylindrical path parallel to the axis of rotation of the rotor blades, the path being progressively reduced in cross-sectional area, to cause acceleration or compression of the air according to circumstances. The axial flow compressor is nowadays familiar in the aviation gas turbine, in which it rapidly supplanted the centrifugal compressor at an early stage in the history of jet-powered flight. It is a commonplace observation that sharp changes of direction of air are not conducive to efficiency of flow; and it has also been observed that airflow through a centrifugal compressor cannot help but be turbulent. By contrast, the airscrew employed for aircraft propulsion is a notably efficient device, commonly attaining an efficiency of 85%; and indeed it is said to be difficult to make one less than 70% efficient. It ought therefore to follow that if the same principles were applied to a compressor, a corresponding efficiency might be obtained; and the theories developed for aerodynamic work on airscrews have therefore been applied to the axial flow compressor with good results. Indeed it is in the matter of adiabatic efficiency that the main claim for this type of compressor is founded, values of 85 to 90% being usual. Against this must be set the disadvantage of bulk and weight being both greater than in the centrifugal machine, while the useful range of flow is narrow and the danger of surge much greater.

There have in fact been some small axial-flow compressors made in the USA for supercharging car engines, but they were not a success. Looking back at all the various types of compressors that have been or could be applied to the car engine, it seems that there are only two that deserve serious consideration — one is the centrifugal machine, where very high power is sought, the other is the Roots positive displacement blower, suitable for quite low boost pressures, in applications where the rather poor adiabatic efficiency of this supercharger would not constitute a serious objection. In principle, therefore, it ought to be a matter of Roots type for roadwork and centrifugal for racing. As it happens the choice is not quite Hobson's, for there is much that can be achieved by multi-stage supercharging with Roots machines set in series, to achieve respectably high performance and efficiency — and on the other hand, where turbocharging is applied to road cars, with the object of ensuring not only high performance but also good thermal efficiency (or economy, which amounts to the same thing) and good manners (which means flexibility, low noise levels and clean exhaust), the centrifugal blower is the invariable choice for reasons to be explained in Chapter 7. Nor is the choice unlikely to be extended: there is still scope for further development of the Wankel engine as a high-performance supercharger, while the tangential fan — already familiar in domestic and industrial apparatus — presents attractive possibilities, too.

Chapter 4
The fuel system

If compressing the charge air makes it hot, the introduction of fuel can cool it. The extent of the cooling depends on two things: one is the point at which the fuel is introduced, the other — and much more significant — is the latent heat of evaporation of the chosen fuel.

Provided that the first condition is satisfied, giving the evaporation of the fuel time to cool the air before it reaches the cylinder — a question to which we shall return later in this chapter — the effect of fuel cooling can be valuable. If air alone is passed through a blower of about 60% adiabatic efficiency, to achieve a boost pressure of 1.5 atm, the delivery temperature will be in the region of 80^0C compared with an ambient temperature of 20^0C. With petrol added, the delivery temperature will be brought down to about 60^0C; and if instead of petrol the fuel be an alcohol mixture, the charge can quite feasibly be reduced to the ambient temperature, or even below. This is clearly equivalent to an improvement in the adiabatic efficiency of the supercharger, and therefore may be equated to a reduction in the power required to drive it; furthermore, the weight of charge inspired by the cylinder is greater because of the greater density of the cool air, and will be accompanied by safer and more efficient combustion due to a reduction in temperature of internal danger areas such as the exhaust valves and spark plugs.

It is common knowledge that the use of alcohol fuels permits a much higher compression ratio than does petrol. They are usually based on methyl alcohol or methanol, CH_3OH, which has a much lower calorific value than petrol — in other words from a given amount

43

of the liquid there is less heat to be obtained from it — and the internal combustion engine is, when reduced to basic principles, a heat engine. The more heat you can use by burning fuel in it, the more power will it generate; and a pound of methanol will contain only about half as much calorific value as a pound of petrol. On the other hand, alcohol can be burned in a much richer mixture, in a piston engine, than can petrol. The enrichment of a petrol/air mixture beyond 1:12 leads to a substantial loss in power, whereas the engine can tolerate alcohol mixtures as rich as 1:4 (by weight) without power loss. This greater proportion of alcohol in the mixture very approximately compensates for the low calorific value of the fuel; and in a given set of circumstances, the power produced when an engine is running on alcohol may be much the same as when running on petrol. This is a fair measure of the useful work which is available from different fuels: thus methanol, at a maximum-power mixture strength of 5½ parts of air to 1 of fuel, produces much the same heat as petrol at a strength of 12 parts of air to 1 of fuel. This of course applies to engines which are the same in all respects except in mixture control; but with alcohol a much higher compression ratio can be used, and therefore proportionately more work can be obtained from the engine, if the compression ratio or compression pressure be adjusted accordingly.

The reason for the ability of alcohol to withstand higher compression ratios is its great cooling effect, combined with a naturally high anti-knock value. Dealing with the latter first, it can be said that alcohol has an octane rating considerably higher than 100, and therefore appreciably higher than the petrol generally available. As far as the cooling is concerned, the significant factor is the great latent heat of evaporation of alcohol compared with petrol. In fact evaporating an ounce of alcohol in air means giving up to the air three times as much heat as the amount yielded by an ounce of petrol. Actually the overall cooling effect is six times as great with alcohol as with petrol, because not only is the alcohol three times better in this respect to start with, but also it is used in a mixture which is twice as rich. In practical terms, this means that by the time the piston comes to top dead centre on the compression stroke, the temperature of the charge will be as much as 200°C lower with alcohol than with petrol.

All this is true of an unsupercharged engine, but it applies even more compellingly in the case of the supercharged one. The introduction of liberal quantities of alcohol into the air passing through a supercharging system has a valuable refrigerating effect, upon which the designer of a highly supercharged engine is forced to rely: it is when using petrol that troubles are encountered if comparable performance be sought. Unless superchargers of exceptionally high adiabatic efficiency are combined with elaborate intercooling systems to lower the charge temperature and increase its density before it reaches the cylinders, a petrol-burning engine cannot safely be subjected to a high degree of supercharge. Alcohol is an effective substitute for all this mechanical refinement, lowering temperatures throughout the cycle and particularly during the combustion phase, so that piston crown and combustion chamber walls run cooler, the valve temperatures are lower, and to a lesser extent the engine as a whole runs cooler; which is another way of saying that it can be worked harder before bringing it to the same critical temperature beyond which it begins to turn nasty.

Simple Ford installation of turbocharger on 3-litre Essex engine

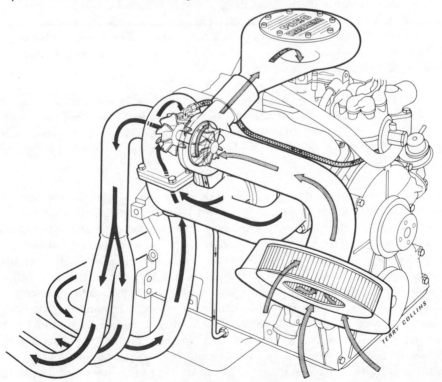

A very simple turbocharger installation by Ford on their V6, blowing into the carburettor. Note the oil pipes to and from the turbine shaft bearing

Turbocharging and Supercharging for Maximum Power and Torque

Apart from giving cooler running, and enabling a higher compression ratio or boost ratio to be used, and therefore yielding more power, alcohol has other advantages. One is that the rate of combustion is slower than that of petrol: instead of a very rapid burning which verges upon detonation, the inflammation of the charge progresses smoothly. So the sudden loadings upon the inward parts of the engine are less, the engine runs more sweetly, and, all other things being equal, will run longer.

Part of the reason for the very rich mixtures used when burning alcohol is the fact that the alcohol is not all fuel; some of it is actually oxygen. Like the fuels used in rockets, alcohol carries some of its own oxygen for burning. This oxygen is released during the combustion process, and by its presence enables more fuel to be burned than would otherwise be the case. Then of course more fuel has to be available if the oxygen is not to be wasted. This idea can be taken a stage further, by using a fuel containing even higher amounts of oxygen; and in rich mixtures this can make a considerable difference, for the atmosphere itself is only 20.8% oxygen. The easiest oxygen-containing fuels for use in the piston engine are the nitro variety, which were in use before the war but remained popular in Grand Prix racing right up to the end of 1957, and are extensively used in drag racing or sprinting today. All these nitros — the class includes nitro-benzine, nitro-methane, and nitrous oxide or laughing gas (which is rather difficult to use) — are chemically related to nitric acid, and they must be employed with extreme caution, because an error in mixture strength can lead either to hideous corrosion of the engine's internals, or to a violent explosion in the fuel tank. Such additives are not for nothing called 'liquid dynamite'!

These oxygen-rich fuels are invariably used in combination with a proportion of acetone, a very useful substance which makes the blend more volatile and thus eases starting problems, while at the same time reducing any tendency to pre-ignition in hot-running engines. A fuel which is basically 80% methanol might have for example 15% acetone, and then a final 5% of nitro-benzine.

There are some practical difficulties of which to beware. Scores of years ago, the London General Omnibus Company carried out tests to determine the best and most economical fuel for their bus engines, and found very satisfactory results with a mixture of alcohol and benzole; but the tests revealed problems with corrosion of ferrous or copper fuel tanks, when alcohol mixtures were used. No appreciable corrosive effects were observed with brass, zinc, tin or aluminium; and lead, for what it is worth, was found completely immune. The design of the fuel system must bear these corrosive susceptibilities in mind; the choice of lubricating oil must be similarly eclectic, for alcohol is not only an effective degreasing fluid, tending to remove the oil film which is or should be present on the cylinder walls and elsewhere, but also reacts detrimentally wtih certain additives commonly present in motor oils on general sale. Most of these contain in their additive specifications (which can amount to a quarter of the total volume of the 'oil') an anti-scuff agent known familiarly as ZDDP (Zinc dialkyl dithiophosphate) the action of which under extreme pressure is vital to the life of certain components

Front view of BMW 2002 Turbo charger

subjected to severe surface stresses, notably camshaft followers and tappets, and valve rocker tips. In the conditions prevailing within a working engine, ZDDP reacts with alcohol to form an abrasive compound with a singular propensity for engine wrecking.

These practical difficulties are of small concern compared with what might be described as social ones. Alcohol fuels are quite out of the question for road cars with any claim to autonomy, for they are simply not available at roadside filling stations. Nor are they permitted by the rules governing most types of racing; only drag racing, sprints and a few very specialised classes of track racing in the USA, and vintage or historic car racing in Britain, constituting exceptions. For all other purposes the choice is Hobson's: petrol or nothing. As we have seen, petrol imposes severe limitations on the extent to which supercharging can be carried out, because of its sensitivity to knock, and its relatively low latent heat of evaporation. If it cannot be relied upon to do very much in the way of charge cooling, then that cooling must be supplied by other means; and of these we must note particularly the device known as the intercooler, sometimes confusingly known as an aftercooler, or more reasonably as a charge air cooler.

Inlet side of the BMW 2002 Turbo engine

An intercooler is simply a heat exchanger inserted in the induction circuit at some convenient point between the compressor outlet and the cylinder inlet ports. The heat exchanger may use air, water or some other low-temperature fluid as the cooling medium. The degree of cooling which may be accomplished is a function of the temperature of the cooling medium, the size of the heat exchanger, and the rate at which the medium can be circulated through it. The unattainable ideal performance of an intercooler would be to bring the temperature of the charge air down to that of the cooling medium; the actual temperature drop of the charge compared with this is a measure of heat exchanger efficiency. A fair figure for this is 70%, and this enables us to assess the relative merits of the different cooling media available.

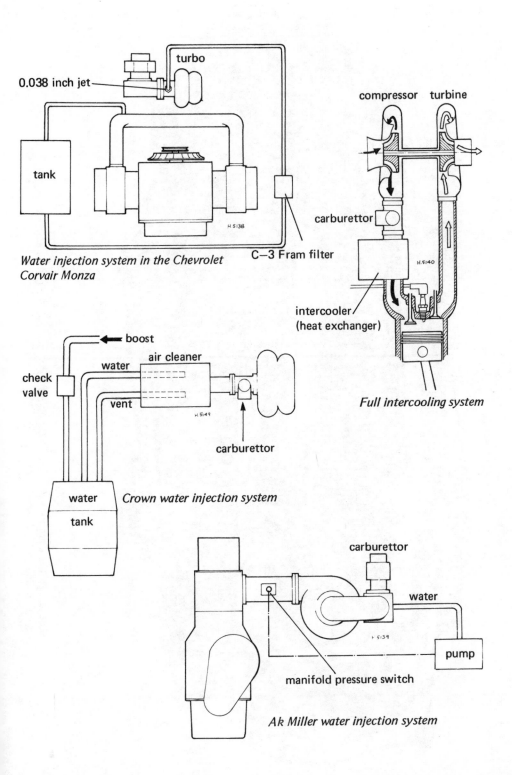

0.038 inch jet

turbo

tank

Water injection system in the Chevrolet Corvair Monza

C–3 Fram filter

compressor   turbine

carburettor

intercooler
(heat exchanger)

Full intercooling system

boost

check
valve

water

air cleaner

vent

carburettor

water
tank

Crown water injection system

carburettor

water

pump

manifold pressure switch

Ak Miller water injection system

49

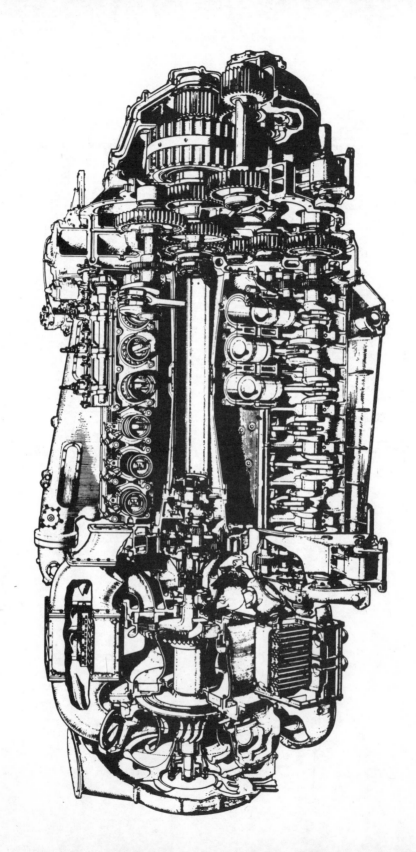

The turbo-blower aggregate of the charge-cooled Napier Deltic is nearly as big as the three banks of six cylinders with their total of 36 opposed pistons

Air is justifiably the most popular. It costs nothing, and is always available; and although to achieve a given degree of cooling one needs four times the weight and 4,000 times the volume of air as of water, it can be picked up and discarded as the car goes along, albeit at some slight cost in aerodynamic drag. Water, unless used in an entirely separate system which will demand its own water heat exchanger during a run of any appreciable duration, will have to be taken from the engine cooling circuit, and the heat given up to it in the supercharging intercooler will have to be dispersed by the engine cooling radiator, so the same losses will still be present.

Let us then consider some examples based again on our previously postulated supercharger delivering air at a temperature of 70°C above the ambient 20°C. The interpolation of a 70% efficient air-to-air heat exchanger will reduce the charge air temperature by 49°C, leaving it at 41°C on entry to the cylinders. Using an engine jacket water intercooler, no useful cooling could be achieved: the modern high-performance engine necessarily and desirably runs at a fairly high temperature, for the sake of thermal efficiency, and many current cooling systems are designed to function at or about 100°C, being pressurised accordingly so as to raise the boiling point of water as high as 127°C.

A sprint vehicle can carry enough static low-temperature liquid to cool the charge during the few brief seconds that the engine is under load. The reservoir of intercooler liquid might contain iced water or acetone or alcohol, perhaps with pieces of dry ice or solid carbon dioxide added. It is not desirable that the cooling medium should be as cold as 0°C, because of the danger of icing of the inlet with naturally airborne moisture. This is perhaps a more theoretical than real danger, and if we postulate an air-to-liquid intercooler with a medium temperature of 0°C and a 70% efficiency, we find a reduction in charge temperature from 90 to 27°C, which is comfortably close to ambient temperature, and approximately equivalent to what can be achieved by the exploitation of alcohol fuel. Charge cooling can seldom be maintained at such an effective level for more than a minute or two, but so long as it can, it is a very effective remedy for the ills of knock-sensitivity and compression ratio abasement, restoring much of the thermal efficiency commonly lost in supercharging. Account must be taken, however, of the weight and bulk of the intercooler system, and of such power as may be necessary to induce a flow of air through the heat exchanger matrix.

The really efficient intercooler can have some curious effects on a turbocharged engine. Remembering that a drop of temperature in the charge will be reflected by a similar drop in the exhaust temperature, it will be seen that the turbine inlet temperature will be reduced, and the turbocharger will slow down, reducing intake manifold pressure. This can be countered by using a smaller turbine housing to maintain the same boost pressure, using the matching technique to be described in Chapter 7.

The practical objections to the intercooler are based on its weight and particularly upon its most inconvenient bulk. There is another system of intercooling which can be much more conveniently installed, and need not impose much weight penalty, and this is water injection. The introduction of water to the charge air delivered by the supercharger has the same cooling effect as the introduction of fuel, and when the water vapour is changed to steam in the combustion process, it acts as a detonation suppressant, possibly even augmenting the power if in the ensuing chain reactions the oxygen and hydrogen atoms of the molecule are separated to increase the combustible content. On the other hand, it is generally thought that if water is introduced in conditions where detonation is not a danger, its effect is to reduce power output. In unsupercharged engines, water injection has often been tried as a means of gaining either power or economy, or even permitting the use of low-octane petrol, but it has always been difficult to arrange for the water dosage to increase in proportion to the combustion pressure or temperature. In supercharged engines it is much more easy to arrange, so that it can be self regulating: the water delivery is simply made proportional to boost pressure by connecting the inlet manifold to the water reservoir, perhaps with a spring-loaded valve on the water delivery line, set to cut off the flow whenever boost pressure falls below some arbitrarily chosen level which might be in the region of 1.3 atm.

Obviously the water reservoir and plumbing must be sufficiently robust to withstand whatever pressure the supercharger delivers; equally obviously the quantity of water to be introduced can only be ascertained empirically, and the size of the reservoir will have to be calculated accordingly if the device is to be run for long periods. For any kind of steady-state operation, the storage problems involved might be more difficult than those created by an air-to-air intercooler; but for brief bursts of knock-free power, water injection is a valuable adjunct to a supercharger, especially if the fuel be petrol. Where there are some regulations prohibiting the use of alcohol, the water must be used neat; where there are not, it may be laced with alcohol in a proportion not exceeding 50%. The alcohol should be methyl alcohol, because of its latent heat of evaporation, for the principal object is to increase the volatility of the mixture, and therefore its cooling effect, rather than to add fuel. A further benefit conferred by the alcohol content, and one of overriding importance in the aviation installations where the technique was first developed, is to act as an anti-freeze agent.

The introduction of a cooling fluid to the compressed air in the inlet manifold is analogous to the introduction of fuel, and so brings us to the question of where this should be done. Prior to the introduction of fuel injection, which eliminates some of the problems, this was resolved into a question of whether the supercharger should blow or suck through the carburettor. This raises two, sometimes three, problems for study: first, the provision of efficient carburetion, secondly, the limitation of temperature rise, and in some cases thirdly, the adaptation of an existing engine to supercharging.

It is commonly this third problem that tempts engineers to retain an existing carburettor installation and supply it with pressurised air from the supercharger. What the carburettor cannot effectively do then is to compensate for the increased density of the air passing through it. The carburettor conventionally works from an atmosphere of substantially constant density, and regulates the supply of fuel according to the velocity of that air

Manifolding of the BMW Turbo delivers air from the turbocharger (invisible beneath exhaust manifold on right) to the airflow metering unit of the Bosch injection system and thence to the four cylinders. Fuel injection takes place immediately upstream of each inlet valve

through the venturi, where its pressure is reduced, the pressure differential being used to draw fuel through a type of manometer. When the air is densified by mechanical compression prior to its introduction to the carburettor, it flows through the venturi at a lower velocity; and although the necessary internal pressurisation of the carburettor may compensate for the weakening effect thus introduced, it does not do so exactly, and thus it becomes difficult to ensure the provision of correct mixture strengths over the range of manifold pressures extending from negative to positive boost.

Internal pressurisation of the float chamber in the carburettor is necessary, because without it the air pressure supplied by the supercharger would simply blow fuel back through the jets, and out of the float-chamber vent holes. Instead, the float chamber must be linked by a balance pipe to the inlet manifold so that the equilibrium in pressure of air supply and fuel supply is restored. This creates a number of practical engineering difficulties, including the effective sealing of the carburettor from the atmosphere. Throttle spindles and the like can be sealed by O-rings or labyrinth seals at some inconvenience in machining; float chamber and other gaskets must be pressure-tight; and

53

the float itself must be made of a material that will not permit its collapse when subjected to external pressure. The carburettors of modern engines hung about with all the panoply of emission controls, such as have become mandatory since 1970, are even more difficult to pressurise, especially if they are piped for the control of evaporative losses: whether petrol vapour be vented to a charcoal canister, to the inside of the air cleaner, or direct to the tank, makes little difference to the nature of the problem.

Some carburettors are almost impossible in the blow-through application, notably those of the constant-vacuum type such as the SU and the Stromberg. The fixed-choke types (Carter, Holley, Solex and Weber are the most common) are more amenable. There is also a very small number of carburettors which work perfectly well in the blow-through mode, the best known in car applications being the Fish, in which the fuel supply and metering work on somewhat different principles from conventional carburettors and can function without regard to the density of the air supply. The most notorious shortcoming of the Fish carburettor, which is in many ways extremely efficient, is its tendency to icing; this is due to the unusually effective atomisation of the fuel discharged into the airstream, which has a pronounced cooling effect. The refrigeration thus induced in the carburettor body encourages airborne moisture to freeze on the internal surfaces of the carburettor, and the resulting accumulation of ice can seriously interfere with air flow and mixture strength. With other carburettors — and all types are prone to icing in certain conditions, usually when atmospheric humidity is high and temperature a little above freezing, say, 5°C or thereabouts — the danger might be avoided by the provision of some kind of artificial heating. In the case of the Fish carburettor, since it functions so well in the blow-through mode, it is an attractive proposition to rely on the high temperature of supercharger air delivery to forestall any risk of icing.

Virtually all automotive carburettors (the rare exceptions are the floatless types which have not been in production since the early 1960s) embody a float chamber to regulate the level of fuel in relation to the jets and control the supply from the fuel pump. When the float chamber is pressurised, to balance the manifold pressure when the carburettor is mounted downstream from the supercharger, there is a danger that the pressure within it will exceed the delivery pressure of the fuel feed pump and no fuel will be available to flow into the chamber. This could easily be overcome by specifying a pump of greater delivery pressure than the average, which is only about 1.3 atm; indeed it is quite possible to find fuel pumps capable of sustaining delivery pressures far higher than any conceivable boost pressure. However, that is not an end of the matter: whenever the engine is running at part load or on the overrun the boost pressure will drop, and the high-pressure fuel pump, denied its opposition, will promptly flood the float chamber with far more fuel than it can handle.

There are ways around the difficulty. One is to use a supplementary electric high-pressure pump, brought into action by a pressure switch mounted on the manifold, whenever the boost pressure rises above the critical level. A more reliable and self-regulating system, which works without the jerky on/off delivery characteristic of this

Upright Roots blower of C-type GP Auto-Union fed a plenum chamber integral with the castings of the V16 engine, saving the weight of a separate manifold but sacrificing all hopes of charge cooling

electric system, can be arranged using one of those mechanical fuel pumps in which the spring chamber is, or can be sealed from the crankcase. The Volkswagen fuel pump is one such that is readily available: after ensuring that the pushrod which communicates engine eccentric-cam motion to the rocker arm is suitably sealed in its guide, the lower chamber of the pump may be pressurised by connecting a balance pipe from it to the inlet manifold, so that the spring will be assisted by the boost pressure, and will always deliver standard pump pressure plus boost pressure. The pump thus becomes its own pressure regulator, always delivering fuel at a pressure that is adequate but never more. An additional attraction of the system is that the boost pressure will help prevent the pushrod sticking at the top of its stroke, an operational fault that makes the pump stop working, and may be aggravated by the necessarily snug fit of the pushrod in its guide.

The same problem of contra pressure interfering with fuel delivery applies theoretically to fuel injection systems, in which it is necessary that the fuel pressure at the nozzle should be substantially higher than any forseeable boost pressure if flow through them is to be satisfactory. In fact this is a problem that is more theoretical than real, for nearly all fuel injection systems, even those described as of low-pressure type, operate at a pressure level higher than is practical for the boost pressure of a petrol-burning engine. In virtually every other respect, an injection system is free from the troubles affecting a carburettor in the blow-through mode, and the injector nozzle or nozzles may be located virtually anywhere in the intake system without fear of the mechanical consequences. The effect of injector location on thermal efficiency and mixture distribution is rather more critical:

if the nozzles are in the inlet ports, as is the case with most unsupercharged engines boasting petrol injection, the temptation to leave them there when adding a supercharger or turbocharger may be very strong — so strong indeed that even BMW and Porsche have succumbed to it. It must be obvious, however, that to introduce the fuel to the charge air so late is to give it virtually no opportunity to have any cooling effect, whereas if the port injectors were replaced by a single continuous-spray nozzle injecting fuel into the air at the compressor outlet, then evaporative cooling could take place throughout the length of the inlet manifold. Going further, it can be argued that since the supercharger, whatever type it is, does not achieve truly adiabatic compression, even better results would be obtained if, instead of cooling the compressor discharge air, it were prevented from getting so hot in the first place; in other words, the injector nozzle should discharge into the airstream on the suction side of the compressor. The same reasoning prompts the location of a carburettor there on the suction side rather than on the pressure side; and in this location the carburettor will be able to function perfectly normally as its designers intended, without any problems of fuel pressure or air density. Indeed the carburettor may function more efficiently when subjected to the steady suction of a supercharger than it can when affected by the intermittent pulsations of an unsupercharged engine's inlet manifold; and this circumstance gives the constant-vacuum type of carburettor a particular claim to preference as the most effective carburettor for supplying mixture to a supercharger of any sort. Indeed, bearing in mind the distinction that when working normally a carburettor gives the engine what it wants, whereas a fuel injection system gives it what the designer thinks it should have, and further bearing in mind that the constrictions and obstructions presented by a carburettor to air flow into an unsupercharged engine need have no detrimental effect upon the volumetric efficiency of a supercharged one, it is very difficult to make out a convincing case for the use of fuel injection apparatus in conjunction with forced induction.

An exception might be made in the case of a racing engine imbibing very rich alcohol mixtures. Here — and even more so if oxygen-bearing additives of the nitro family are employed — the mixture may be so wet that there will be problems of fuel deposition on the walls of the compressor (due to its centrifuging action), despite the common supposition that the action of a supercharger includes some mechanical stirring of the mixture to improve the distribution of droplets evenly throughout the charge air. In such a case, it is more desirable to use some form of injection system at the compressor outlet; it can be a very simple one, because of the freedom that alcohol gives from the disciplines of mixture control; relatively crude devices such as the old Hilborn fuel injection apparatus can be used with impunity, though it is usually thought advisable to incorporate a compensator that automatically increases fuel pressure to the nozzle as boost pressure increases. Probably the ideal compromise is that chosen by some American racing engine specialists, who locate an idling fuel nozzle in the intake eye of the supercharger (which is invariably of the centrifugal type associated with an exhaust turbine) so as to make the most of the evaporative charge cooling that this basic fuel supply can provide. If the first rule of supercharging must be to avoid being too greedy for power and high boost pressures, the second must be that charge cooling cannot be taken too seriously.

Chapter 5
Inlet manifolding

There is no doubt that supercharging is expensive, not only in first cost but also in subsequent fuel bills. In exchange one buys freedom from a large number of constraints not only in driving but also in design: and both points of view are relevant to the shape and layout of the inlet manifold — which is strictly the conduit between the compressor outlet and the engine inlet ports, though we may stretch a point and include the air ducting to the carburettor and/or compressor inlet.

The point is made in the first chapter that the highly tuned unsupercharged engine is inherently inflexible, demanding operation within a fairly narrow speed range if much performance is to be realised. These confines are generously extended by a good forced induction system which, by guaranteeing high volumetric efficiency at all material times, makes it quite unnecessary to go in for streamlined induction tracts of painstakingly calculated and experimentally verified (or, as is more likely, altered!) length to secure resonance at some chosen frequency. In fact both resonance and streamlining are undesirable in a supercharged system.

With Roots blowers of the simplest type, some pulsation in the air delivery is difficult to avoid, but it can be effectively damped in the downstream manifolding. Whatever the type of compressor, undamped resonant waves issuing from the inlet ports as the valves open and close could have a detrimental effect upon the air flow within the blower, and thus upon its performance. In any case, resonance will tend, as in an unsupercharged engine, to interfere with cylinder filling at certain speeds, which can make a nonsense of the ignition advance curve if it does nothing worse.

When Daimler-Benz moved the carburettor of the W125 Mercedes-Benz racing car up-stream of the supercharger, they also adopted this simple but efficient inlet manifold gallery featuring more turbulent flow than in earlier, more 'streamlined', versions

One of the worst things it can do is to interfere with the working of the pressure relief or blow-off valve that is a desirable addition to the inlet manifold. This relief valve is not intended to control maximum boost pressure, and would be deplorably wasteful if it were so used: in practice it is usually set to open when the inlet manifold pressure is

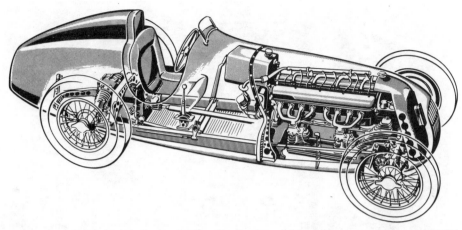

Observe the cold-air intake (penetrating radiator) for the blower of the 1934/5 GP Mercédès-Benz

perhaps a quarter of an atmosphere above the designed maximum boost pressure. When it does, it is to guard against the danger of the system being blown apart by a transient high-pressure pulse, such as might be caused by a blow-back from a cylinder when its inlet valve is open. The point about resonance is that these spring-loaded relief valves are in effect very short resonance chambers with a natural frequency that may be several times that of the highest fundamental frequency likely to be reached in the inlet manifold — but the spring of the valve could very easily be induced to vibrate in resonance with one of the higher harmonics of what might have been thought a safely low fundamental oscillation in the manifold, and the effect of this misbehaviour could be disastrous. It is therefore doubly desirable to inhibit any tendency to resonant pulsation in the manifold, and it is good design practice to locate the blow-off valve or valves where there is least danger of them being triggered open by spurious vibration.

As to streamlined flow, so necessary in the unsupercharged engine in order to secure reasonably high volumetric efficiency, it is quite unnecessary in a supercharged system, and in fact can be detrimental to performance, because laminar air flow can cause swirl and induction bias, resulting in some cylinders getting more than their proper share of mixture, and others less. Since streamlined flow is so very difficult to ensure, its undesirability becomes a positive boon to the designer, who need do little more than provide piping of large cross-sectional area — when we have plenty of pressure we have no need of velocity— and ensure turbulent flow of the air within it, which is all too easy. A couple of sharp turns in the pipework will do it, a heat exchanger will do it, as would some deliberately introduced vortex generator; there have even been cases of the manifold being baffled by a transverse plate with holes drilled in it to pass the air in a turbulent flow. It is another reason for eschewing streamlined curves from the feed pipe to the individual inlet ports of the engine: it is far better for the branch pipes to be perpendicular to the main pipe, and with a view to minimising resonance that they be as short as

Inlet side of the 1934 GP Mercédès-Benz engine reveals pressurized carburettors and phased manifolding, both of which were later proved unsound

possible. Nor is there any need for the main pipe — which is really a plenum chamber — to be of circular cross-section; there is a lot to be said for making it of rectangular section, not only because it eases fabrication but also because it increases the surface area in relation to the volume of the manifold, improving its heat dissipation and thus helping charge cooling. In cases where a rich wet alcohol mixture is being passed, the increased surface area provides a greater opportunity for the internal walls of the manifold to act as a kind of surface carburettor, the deposited fuel being drawn off the surface by evaporation.

Only in two places is any attempt at streamlining of the airflow justified. One is upstream of the compressor inlet, where the air flow is similar to, albeit more continuous than, that in the induction system of an unsupercharged engine. Air enters this section at ordinary atmospheric pressure, and all the rules for attaining high volumetric efficiency through it still apply: the entry should be through a flared convergent bell or nozzle (the ideal shape is that of an exponential horn, which resonates equally at all frequencies but is an inefficient reflector of waves) and if it then passes through a carburettor, the internals of that instrument should be streamlined by whatever techniques are appropriate to its type, such as by the use of lenticular butterfly throttles, streamlined jet holders or secondary venturi supports, and so on. Because it is passing air at atmospheric pressure, the carburettor will have to be of relatively large bore: carburettor manufacturers can quote the air-swallowing capacity of their instruments, usually doing so in terms of cubic feet per minute. It should be noted that the bore of a carburettor located downstream from the compressor should be appreciably smaller, because of the greater density of the air passing through it.

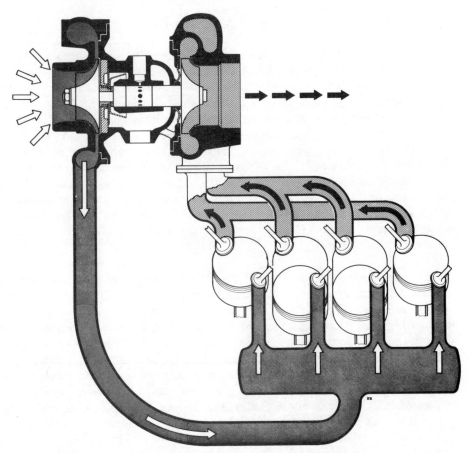

Schematic layout of turbocharger feeding air to the BMW 2002 engine; the fuel injection nozzles in the inlet ports are not shown

Downstream from the compressor there is one place where streamlining is perhaps desirable, this being the junction of the main manifold plenum chamber and each of the individual pipes communicating with the engine cylinder intake ports. This junction should have a radiused edge, because a sharp edge would cause stalling of the air flow across the orifice, possibly to a different extent at each pipe entry, so that they would not all be choked by the same amount. The result would be a distribution bias in which some cylinders were dosed with less charge than others.

The danger of introducing distribution bias also occurs where the inlet manifold plenum chamber is supplied by more than one compressor discharge. Even though the conduit from the compressors simply dumps its charge straight into a rectilinear box so as to encourage the turbulence sought, there is a risk of a swirl pattern being promoted if the two tributary flows are offset — and it is swirl that causes most maldistribution problems in a supercharged induction system. The answer is to ensure that the two tributary flows enter the plenum chamber in direct coaxial opposition.

Turbocharging and Supercharging for Maximum Power and Torque

The problem is one that is only commonly encountered in turbocharged V engines, (usually large-capacity American V8s) on which it is convenient for a number of reasons for each bank of cylinders to exhaust into its own turbine. It is not unusual in these cases for a conventional V8 inlet manifold, one of those complex castings full of tortuous passages, to be retained for mixture distribution from the plenum chamber, in which case there is a further danger of distribution bias arising from the sheer performance of the vehicle. This is a phenomenon also noticed in drag racers employing a mechanically driven supercharger drawing from injector bodies or special suction carburettors: the acceleration of these vehicles is so great that the fuel droplets entrained in the charge air tend, because of their inertia, to be left behind by the vehicle, the effect being to supply an excessive mixture to the rearward cylinders while those at the front are starved. A more complicated plenum chamber, or careful positioning and sizing of the injector nozzles or jets, can offset this bias.

There remains one more thing to be said about the inlet system, about the point where it all starts. An intake air cleaner is so desirable as to be considered virtually essential. The fine dust particles which eventually wear out a touring car engine may not be objectionable in a high-performance car of limited life expectation, but larger foreign bodies can wreak havoc. The ingestion of a small stone, say, or one of the little screws that secure a carburettor butterfly plate to its spindle, may with luck do no more harm than perhaps cause a slight nicking of a valve seat as they pass through the unsupercharged engine; but any such object entering a supercharger will do much more costly damage, and luck will be conspicuously absent from the probabilities. Some kind of gauze or screen may be sufficient to keep out undesirable objects, but a proper paper element air filter is better, especially if a vane type of supercharger is employed. As pointed out in Chapter 3, blowers of this kind require internal lubrication, and if airborne dust mixes with the oil it can form a fine and fatal grinding paste. The chosen air filter must obviously have an air flow capacity at least equal to that of the rest of the system, and since it forms part of the unpressurised section, it should be designed with some thought for streamlined flow. Often the best arrangement is to have a substantially oversized air cleaner serving a large plenum chamber, within which the carburettor or supercharger air intake bell is located.

Chapter 6
The engine-driven compressor

Any of the superchargers discussed in Chapter 3 can be mechanically linked to the crankshafts of an engine for positive drive. According to the type chosen, the requirements of the drive will be rather different: the questions to be considered must be how to drive it, how fast to drive it, and what provisions, if any, to make to forestall any danger of mechanical failure of the drive or of the compressor itself, due to some shortcoming in the drive system. There is plenty of choice of mechanisms: superchargers have been successfully driven by chains, gears, V-belts, toothed belts, torsion bars, friction clutches, fluid couplings or by simple direct connection to the nose of the crankshaft.

The speed at which the compressor is driven must be determined first by the type, and then by its size. Compressors of the vane type are doomed to be run at relatively low speeds, often slower than the engine crankshaft, and the drive system need not embody much in the way of engineering refinement. The only desirable feature might be a pre-set torque limiting clutch of some sort, because the rotors often have considerable inertia, and their vanes may need protection from the consequences of transient shock overloads; and because the out-of-balance forces within the blower should not be capable of being sensed as a vibration at the crankshaft, lest there be some untoward effect on the latter component's sensitivity to torsional flutter. Sufficient cushioning may be provided by a simple V-belt drive, though a duplex belt system is more desirable.

At the other extreme is the centrifugal compressor, which only works effectively at very high speeds. Here the limit is imposed by the tip velocity, which should approach the speed of sound and in some designs is even required to exceed it. Thus, for an engine running up to a maximum crankshaft rate of 7,000 rpm, the requisite limiting impeller tip speed could be achieved with an impeller of six inches diameter, driven at six times crankshaft speed. There would be nothing to gain by doubling the diameter of the impeller and halving the gear multiplication; the inertia of the larger impeller would be about four times greater, demanding much more elaborate cushioning of the drive — and the greater bulk of the compressor could not conceivably be advantageous.

With gear multiplication ratios of this order, simple belt and pulley systems will not work satisfactorily, and multi-stage ones will create difficulties due to the high belt velocities involved. Roller chains may be unsuitable, but a gear drive system may be eminently satisfactory. The necessary gears will add their own inertia to that of the impeller, and any slipping clutch or other torque-limiting device should therefore come between the crankshaft and the gear train, not between the gear train and the supercharger. The use of a long torsion bar to cushion the drive is very popular in aero-engines, and has been seen elsewhere: in the V16 BRM, where the power take-off from the crankshaft was through gears to a half-speed countershaft, the drive to the supercharger passed through a long and slender torsion bar to a torque-limiting clutch built into the first of a series of spur gears, which finally raised the impeller speed to run four times higher than that of the crankshaft, which itself was intended to reach as much as 12,000 rpm. Where part of the drive system, as here, runs at relatively low speeds, it might be possible to substitute for the pre-set overload clutch a free-wheel or sprag clutch, following the argument that the inertia of the impeller is not much of a problem when accelerating from relatively low speeds, but causes difficulties in any attempt to decelerate it rapidly from high speeds. If this course be adopted, care must be taken in selecting the sprag clutch, for some types do not take kindly to sudden engagement and acceleration and have been known to burst when suddenly loaded in this way. Others, developed for various transmission requirements, are immune from this danger.

The blower that gives the most choice of drive systems and speeds is undoubtedly the Roots type. It is capable of running up to very high speeds, being in perfect balance and presenting no frictional problems. As remarked in Chapter 3, the internal leakage is at its worst at low speeds, so a small blower geared to run relatively fast will give more consistent boost pressure over a reasonably wide speed range than a large one run proportionally slower. Of course this can be overdone, with problems of turbulence and stalled airflow around the lobes of the rotors, and at the entry and discharge ports, looming large when the speed becomes excessive. The Roots blower is a positive displacement type, and its delivery at effective speeds can be calculated with reasonable accuracy. It is usual to select a blower or blowers of such a size as will permit them to be driven at a speed somewhat higher than that of the crankshaft but not more than twice as high.

This is a régime in which most kinds of mechanical drive may prove reasonably satisfactory. Torque-limiting clutches are unnecessary, and although some valuable cushioning effect may prompt the use of belt or chain drive, it is perfectly feasible to employ gears of

Monza 1934: after 11 years of successful supercharging there was another 17 years to go before an unsupercharged car would win a major GP. Here the sudden spurt of progress is illustrated from rear to front by Maserati (1.6 atm boost, 167 lb/in² peak-power bmep), Alfa Romeo (1.6 atm, 174 lb/in²) and Mercedes-Benz (1.66 atm, 241 lb/in²)

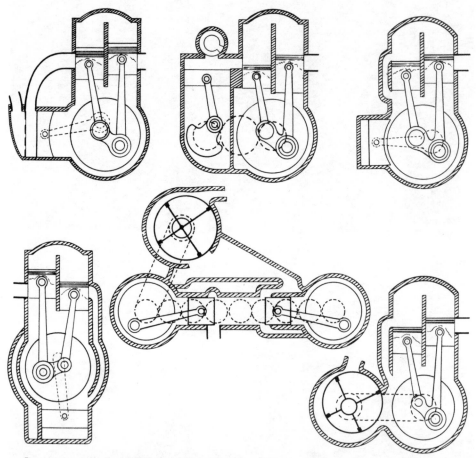

Pressure scavenging of the 2-stroke only becomes supercharging when the exhaust port closes before the inlet. DKW sought such timing with split-single layout in the 1930s: experimental versions included reed and rotary valves and eccentric-vane compressors. The two successful ones relied on short-stroke charging pistons

minimal back-lash, or even direct drive from the crankshaft. From the point of view of sheer mechanical convenience, the toothed belt is undoubtedly the most attractive drive system, especially because it makes any alteration in the crankshaft:blower speed ratio fairly easy. This is something that can be even easier with V-belts, especially if adjustable pulleys are used; but apart from their greater elasticity, which is scarcely needed, they are otherwise inferior to the toothed belt.

If properly sized and driven at the correct speed in relation to the crankshaft, the Roots blower confers reasonable freedom from the problems of speed-sensitive boost pressure that beset the centrifugal blower. Because boost pressure from the latter increases as the square of its rotational speed, it is a real problem to achieve satisfactory boost at much less than maximum engine speed, and several designers in the past have fallen prey to the

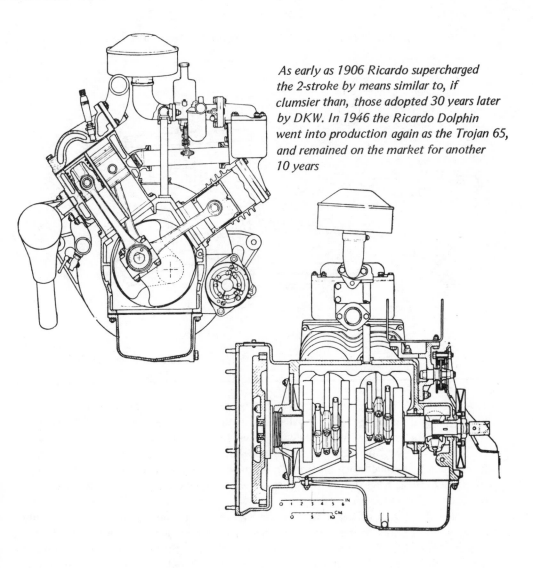

As early as 1906 Ricardo supercharged the 2-stroke by means similar to, if clumsier than, those adopted 30 years later by DKW. In 1946 the Ricardo Dolphin went into production again as the Trojan 65, and remained on the market for another 10 years

temptation to accept dangerously high maximum boost pressures for the sake of reasonable ones lower down the engine speed scale. It seems obvious that some kind of steplessly-variable-ratio drive is wanted, and some American manufacturers of the 1930s did in fact provide it by means of V-belts and adjustable pulleys controlled by a centrifugal governor.

The problem is one that has particularly exercised the aviation industry, because although an aero engine is a sensibly constant-speed device, the lessening of atmospheric density at high altitudes makes it desirable for the blower (almost invariably of the centrifugal type in aero engine practice) to be run at increasingly higher speeds as the aircraft climbs, in order to maintain sea-level power up to as great an altitude as possible. Most of the methods used have been surprisingly crude, involving two- or three-speed gearboxes; but

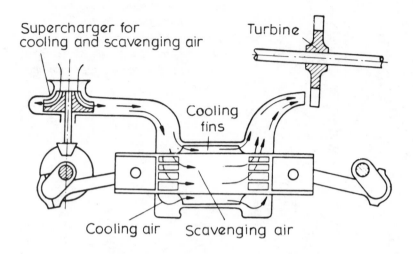

Curious split-delivery pressure air for General Motors' Orion 2-stroke tank engine

a much more refined system was perfected by Daimler-Benz, who hit upon the use of Föttinger fluid couplings in the final stages of the supercharger drive from the step-up gearing to the impeller. What they did was to use two such couplings (miniature equivalents of the so-called 'fluid flywheel' of automobile practice) in series, each independently supplied with oil from an engine-driven pump. One of the couplings was constantly fed with oil to maintain a steady volume of fluid between the two halves of the coupling; the delivery to the other was varied, the quantity of oil in the coupling governing the degree of slip, and thus the overall transmission ratio. Perhaps the only shortcomings of this arrangement were the power losses due to a churning of the oil, and the inevitable slip which is never less than 3% in a fluid coupling of this type. It should perhaps be emphasised that the losses thereby suffered are of power, not of torque — the basic transmission characteristic of a simple fluid coupling is that regardless of the relative speeds of the two elements (the turbine and the pump, as they are called) the input torque is always equal to the output torque.

A similar system might be devised for a car supercharger, though it is doubtful whether the cost of such a development could be justified or even met by anybody likely to be tempted to undertake it. Other types of hydraulic drive have, however, been employed in the past, in applications as varied as diesel trucks and racing versions of production touring cars. One such method is to couple the impeller to a fluid turbine reliant on engine oil pressure which can easily be adjusted to remain substantially constant over virtually the entire working speed range. Another is to use an engine-driven pump with a centrifugal governor reducing its delivery as the engine speed rises.

68

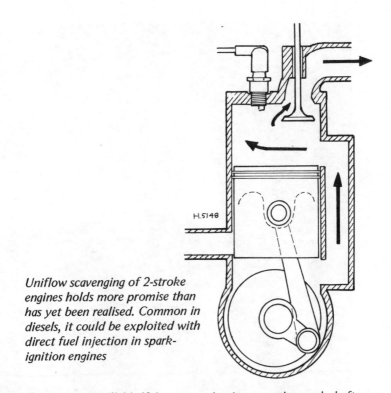

Uniflow scavenging of 2-stroke engines holds more promise than has yet been realised. Common in diesels, it could be exploited with direct fuel injection in spark-ignition engines

H.5148

Much more subtle means of control are available if the connection between the crank shaft and the blower be through differential gearing, but this is a topic that really belongs to Chapter 8. In any case, we must question whether mechanical complication in the blower drive is really necessary in order to achieve the control of delivery from a centrifugal compressor. There are other ways, based on the postulate that if the air delivery is what we really want to control, then it might be better done directly than indirectly. The most efficient is the arrangement known as vortex throttling, sometimes used in exhaust turbines, and proposed by Rolls-Royce for the superchargers of the V16 BRM. This makes use of variable-pitch stator blades or vanes in the diffuser section of the impeller housing: as these vanes are pivoted to alter their angle to the tangential flow of air from the tips of the impeller blades, they alter the cross-sectional area of the diffuser section, making it in effect a variable nozzle. In this way it is feasible to control the boost pressure so that it remains practically constant over a wide speed range. Control of the blade angles may be servo-aided to relate to engine speed or to boost pressure: the former is somewhat easier, but the latter may give the blower slightly greater immunity from the surge that was described in Chapter 3 as being the other bane of compressors that are not of the positive-displacement type.

Because so much was expected of it, so much demanded of it, and so much disappointment occasioned by it, the V16 BRM offers an unusual number of object lessons in supercharging. One more deserves to be mentioned: one of the reasons for its disappointing performance was that the intake air was severely throttled at the cylinder inlet ports. The main reason for this was that the designers refused to countenance high rates

Although the 1938 GP Mercédès-Benz had single-stage supercharging, it had two Roots blowers in parallel feeding a common inlet plenum chamber

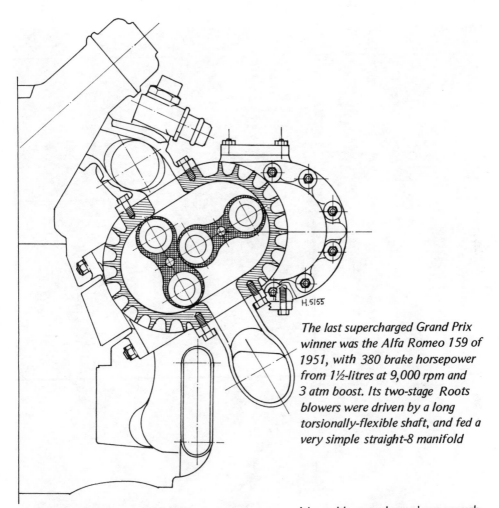

H.5155

The last supercharged Grand Prix winner was the Alfa Romeo 159 of 1951, with 380 brake horsepower from 1½-litres at 9,000 rpm and 3 atm boost. Its two-stage Roots blowers were driven by a long torsionally-flexible shaft, and fed a very simple straight-8 manifold

of valve acceleration, so the inlet ports were open neither wide enough nor long enough. In this connection, it is interesting to note that valve timings vary as much in supercharged engines as in unsupercharged ones, and for similar reasons. Where supercharging is resorted to in order to guarantee ample torque over a wide speed range, as is proper to a road car or a hill-climb car, the camshaft profiles may be gentle and the valve events fairly short, overlap of inlet and exhaust valve openings being significantly curtailed. At the other extreme, in engines required to put out the most extravagant power while consuming richly alcoholic mixtures, wider valve timings with a lot of overlap allow internal temperatures in the cylinders to be kept down, by giving time for a cooling and scavenging blow-through of fresh mixture being blasted in from the blower through the inlet port, scouring the combustion chamber (defined in part by the crown of the piston at top dead centre) and issuing from the exhaust port to be voided in the empty air. It was by such means that the fuel consumption of the type 159 Alfa Romeo came down to somewhat less than 1½ miles per gallon: it was not so much thermal inefficiency as deliberate prodigality — or, to be more fair to the engineers concerned, a kind of insurance against a risk so imminent that the premium was inevitably high.

A very neat two-stage Roots installation in the V8 1½-litre car built by Mercédès-Benz for the 1939 Tripoli GP. Note how small is the second-stage blower, and how large the carburettor

*The 4CLT/48 or San Remo Maserati, here driven by Froilan Gonzales, demonstrated
the value of 2-stage Roots supercharging in making a simple engine competitive*

The same troubles afflict the two-stroke engine which, in most of its forms, suffers from
the fact that the exhaust port always closes after the inlet. Any attempt to supercharge
this type of two-stroke engine is doomed to failure, though there is no doubt that the
scouring and scavenging achieved by a high-pressure blow-through of fresh charge may
be quite beneficial. The only way to convert scavenging into supercharging is to impose
some back pressure on the exhaust system, so that some of the scavenging mixture is
trapped in the cylinder or blown back into it from the exhaust pipe. The best way to
introduce such back pressure is to lead the exhaust into a turbine, which can drive a
blower as the first stage in a two-stage system of which the second is the original
engine-driven scavenge blower. This is, however, by no means the most important
application of the turbocharger, which we will consider in the next chapter.

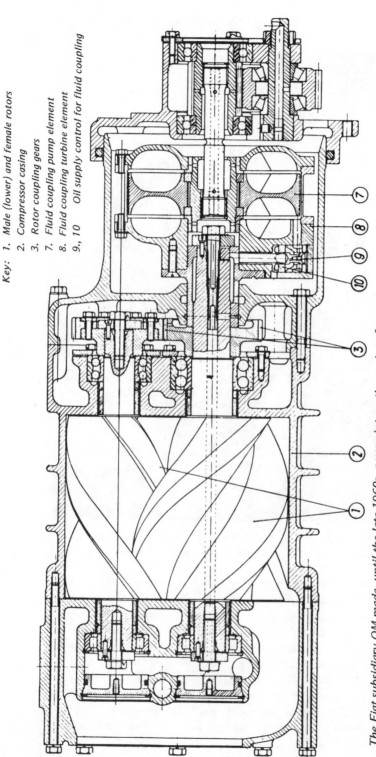

Key: 1. Male (lower) and female rotors
2. Compressor casing
3. Rotor coupling gears
7. Fluid coupling pump element
8. Fluid coupling turbine element
9., 10 Oil supply control for fluid coupling

The Fiat subsidiary OM made, until the late 1960s, a very interesting series of supercharged engines for their DG trucks, which although diesels nevertheless incorporated
many elegant features that might profitably be copied in the supercharging of petrol
engines. The compressor itself was of the screw type, with helical male and female rotors
of three lobes each, imparting a measure of internal compression to what is basically a
Roots configuration and giving it an impressively high adiabatic efficiency. The delivery
of air from this blower could be by-passed in cruise conditions by a poppet valve controlling the air-feed chamber on the end of the compressor (at the extreme left of this
drawing), while the speed and delivery characteristics of the compressor were controlled
by varying the oil delivery to the double-sided Föttinger fluid coupling by which it was driven

A 'top fueler' Miss Revell's blasts off. Roz Prior's dragster again uses a Roots supercharged Chrysler engine. The exhaust gases are directed onto the rear tyres to blow off grit!

Complex by-pass plumbing of the Porsche 911 Turbo engine

Chapter 7
The turbine-
driven compressor

Considering that it is very nearly as old as the engine-driven supercharger, the turbo-charger has taken a surprisingly long time to come into use on car engines. It is begging the question to argue that it was not until comparatively recently that suitably small low-inertia turbines became available: the people who made big ones for trucks and earth movers and aero engines in all the years before would have made them for cars, too, had the demand existed. Nor is it entirely true to explain that it is only nowadays that a concern for fuel consumption has prompted engineers to try turbocharging as a means of achieving reasonably high thermal efficiency together with that other modern phenomen-on, clean exhaust gases. There have been hard times often enough in the past, when fuel economy was sought as earnestly as it is today; but then motorists were content to put up with mediocre performance in exchange for low fuel consumption, whereas modern road conditions, with heavy traffic making brisk acceleration more desirable than ever before, put the motorist in the uncomfortable position of feeling unable to sacrifice either economy or performance. The other reason for the delay in development of the turbocharger is simply that it is a much more difficult business than supercharging with an engine-driven compressor: whereas half a dozen assorted Roots blowers would cope with the requirements of practically all production car engines, the necessary combination of turbines, housings, and blowers to cater for the same range would run into thousands. Hitherto it has simply not been worth anybody's trouble to investigate the problems involved; now the industry is learning very fast, and turbocharging promises to become more fashionable.

Turbocharging of the 1962 Oldsmobile Jetfire was fairly refined but somewhat premature for the market

There are plenty of people who fondly suppose that turbocharging gives them something for nothing. It is not entirely true, for installation can be expensive and it can carry with it sundry operational difficulties. Nevertheless it is attractive because it appears to eliminate the compromise that previously had to be accepted, allowing the volumetric efficiency of the supercharged engine to be combined with the thermal efficiency of the unsupercharged one.

We know from thermodynamics that so long as any mass of gas is at a temperature higher than its surroundings, this gas can be expanded to some lower temperature, doing work in the process. To expand a gas, its pressure must be greater than the pressure into which the gas must be exhausted. Therefore if a temperature and a pressure difference exist between the engine exhaust and the atmosphere, some of the energy in the exhaust gases can be made available.

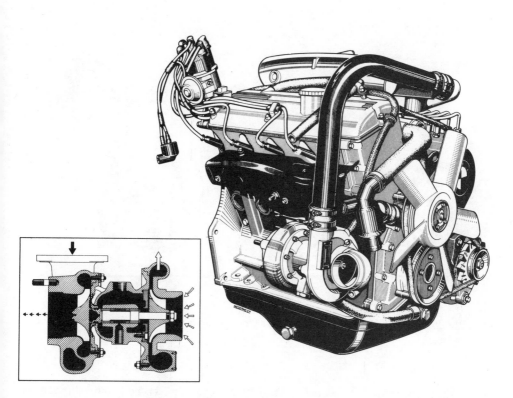

Installation and (inset) cross section of 100,000 rpm KKK turbocharger on the BMW 2002, with petrol injection into the inlet ports

Earlier in this book it was explained that the calorific content of the fuel supplied to a normally aspirated engine was largely squandered: not more than a third would be converted into useful work, the remainder being wasted either in heat shed to the exhaust, or in heat shed through the engine to its coolant or the surrounding air. In a highly supercharged engine the situation is even worse: only about 24% of the potential heat energy in the fuel would be converted into useful power, another 16% would be dissipated through the engine and its cooling system, and the balance of 60% (these are all approximate figures) would be spent down the exhaust pipe. This heat balance looks frightful compared with that of the unsupercharged engine; and it is doubly so if we look at quantities rather than proportions. It means that for a 50% power increase from the same engine, fuel consumption will go up 100% if the power is achieved by simple supercharging.

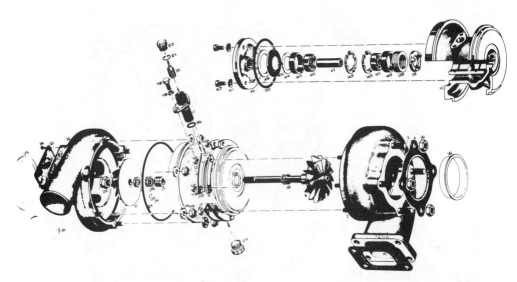

Porsche chose this Eberspacher turbocharger for the already prodigiously powerful 917 sports-racer

From all this we can see that a highly supercharged engine rejects heat to the atmosphere through the exhaust system to the equivalent of about 2½ times the developed power. It is a shocking waste, but according to the first and second laws of thermodynamics there is nothing we can do about it. However, we have seen that, given the necessary temperature and pressure differences between the exhaust pipe and the atmosphere, there is *something* we can do. We know that exhaust gas temperature is much higher than that of the atmosphere, reaching as much as $1,000^0$C. We also know, if only from experience, that any engine will tolerate a certain amount of what we call back-pressure in the exhaust system -- that is, a pressure somewhat higher than atmospheric. So we can deduce that there is some hope of harnessing more of the energy that is in the exhaust gases. This we do by expanding the gases through a turbine, which develops a certain amount of power that can be used to drive the supercharger, or can be fed back into the engine in some other way. Since the total energy that can be made available from the exhaust gases usually exceeds by a considerable margin what the supercharger needs to drive it, it is conceivable that the surplus energy could be recovered in car engines just as it has been in others — but this is for the future, even though that future may be as near as Chapter 8.

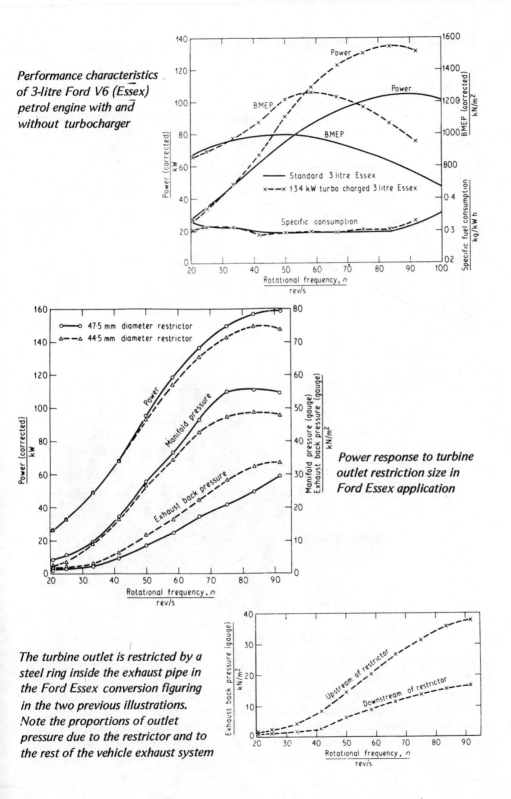

Performance characteristics of 3-litre Ford V6 (Essex) petrol engine with and without turbocharger

Power (corrected) kW

BMEP (corrected) kN/m²

Specific fuel consumption kg/kW h

Power

BMEP

Power

BMEP

——— Standard 3 litre Essex
×–––× 134 kW turbo charged 3 litre Essex

Specific consumption

Rotational frequency, *n* rev/s

Power response to turbine outlet restriction size in Ford Essex application

Power (corrected) kW

Manifold pressure (gauge) Exhaust back pressure (gauge) kN/m²

o——o 47·5 mm diameter restrictor
△–––△ 44·5 mm diameter restrictor

Power

Manifold pressure

Exhaust back pressure

Rotational frequency, *n* rev/s

The turbine outlet is restricted by a steel ring inside the exhaust pipe in the Ford Essex conversion figuring in the two previous illustrations. Note the proportions of outlet pressure due to the restrictor and to the rest of the vehicle exhaust system

Exhaust back pressure (gauge) kN/m²

Upstream of restrictor

Downstream of restrictor

Rotational frequency, *n* rev/s

81

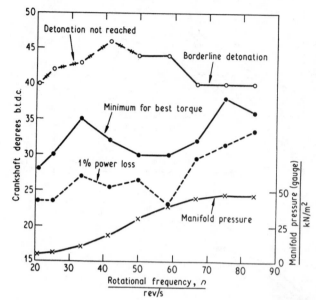

Ignition advance requirements when turbocharging the 3-litre Ford Essex engine

Rotor assembly and bearing housing of Eberspacher turbocharger

In examining the complexities of applying a turbocharger to an engine we may feel some relief in the knowledge that all turbochargers embody centrifugal compressors, the characteristics of which have already been adequately summarised earlier in this book. There is no need to go through it all again; but a promise was given in Chapter 3 that the reason for this universal use of the centrifugal blower in turbocharging would be given here. It is quite simple: only this type of blower can be made sufficiently small and light to have such slight inertia that it can be accelerated as rapidly as turbocharging demands.

Exhaust side of BMW 2002 Turbo engine, showing equal-length 4-into-2 pipes leading to turbine of Eberspacher turbocharger

In the absence of any positive connection between the crankshaft and the impeller shaft, it is possible for the compressor to be turning over relatively slowly, even when the engine is running at high speed, if the driver's foot is off the accelerator pedal and the engine is not under load: conversely it is possible that, with his foot hard down and the engine doing its utmost, the blower may be running at high speed and delivering ample boost even though the engine may be running relatively slowly. When he puts his foot down and opens the throttles he will not be rewarded by an immediate increase in inlet manifold pressure; at least it will be no greater than would be the case in an unsuper- charged engine, where opening the throttles wide raises the inlet manifold pressure from an appreciable vacuum to something approaching atmospheric pressure. Even this will suffice to increase the mass flow of air through the engine, and it is exhaust mass flow that acts on the turbine to accelerate it. The compressor accelerates with the turbine, raising the boost pressure smartly as it does so; and if the response is to be satisfactorily rapid the entire rotating body — turbine, shaft and impeller — must oppose this acceleration with as little inertia as possible. Rotational inertia is a function of mass and radius of gyration (as well as of speed), which may be explained in simple terms by saying that the diameter of the rotor must be kept as small as possible, and most of its material must be concentrated as close as possible to its axis of rotation. This is why turbocharger impellers are even smaller than those of engine-driven centrifugal compressors: both the impeller and the turbine may be as little as three inches in diameter, and although they will be very carefully made and scrupulously balanced, they will weigh only a few ounces.

Because they are so small, and because tip speed must be as high as ever, these impellers and their turbines run up to extraordinarily high speeds, as much as 125,000 rpm. This makes their design a highly skilled business, which is why there are so few firms engaged in it, and why those few must be trusted to give the best available advice on what to use and how to use it. This applies in particular to the thorny subject of sizing and matching.

Sizing is a matter of selecting the compressor and turbine so as to make sure that each can cope with the necessary through-flow; and if you start from the assumption that you know what you want and (which is altogether more likely) that the turbocharger manufacturer knows what he has, it should not be too tricky. *Matching* is a good deal more problematic: it involves selecting the correct turbine housing to ensure that the chosen turbine and impeller can work together to provide the desired boost characteristics throughout the material range of engine speeds.

What that range is depends on the duties forseen for the engine. As already pointed out, what is right for a track racer is quite wrong for a road racer, and so on. It is not only a matter of where in the speed range maximum torque is wanted, but also of how long it will be wanted without respite. Change the circumstances and, although you keep the same engine, you will want a different turbocharger match.

The critical thing here is the proportions of the exhaust turbine housing. This is (at least in the sizes with which we are concerned) usually a simple casting devoid of internal vanes, the only nozzle involved being the passage between the exhaust pipe and the volute housing. What happens within is that the gases pursue a free vortex path, swirling in towards the axis of rotation of the turbine, and gradually changing their direction from tangential, through radial, to axially emergent flow. Two dimensions are critical, or rather the ratio between them: one is the area of the intake throat or nozzle, the other the radius of entry which is the distance from the centroid of the nozzle area to the centre of the vortex. It is the ratio of area to radius that determines the flow. A large-area nozzle working at a small radius allows more gas to flow at a steeper angle in towards the centre, and in these circumstances the turbine rotor will spin more slowly than it would in a housing with a smaller ratio of area to radius. Thus it can be seen that, with a given turbine, changing the housing is comparable to changing the gear ratio or belt puliey

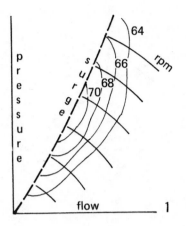

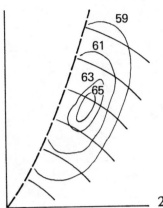

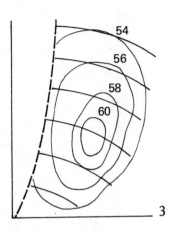

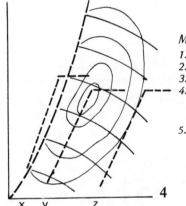

Matching and sizing: clues in the maps

1. narrow island of efficiency, 70% efficient and very accurate sizing required
2. broader island, 65% efficient, care in sizing required
3. very wide island, only 60% efficient but very easy to size
4. x - bad sizing
   y - good sizing
   z - bad sizing
5. x - mis-match
   y - good match
   z - mis-match
6. solid line shows standard power curve
   dotted lines show variations of matching
   x - mis-match, too small an A/R
   y - happy match, compromise
   z - mis-match, too large an A/R

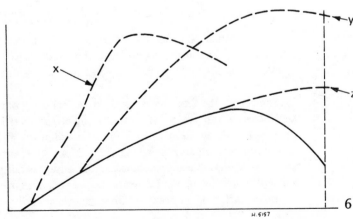

H. 5157

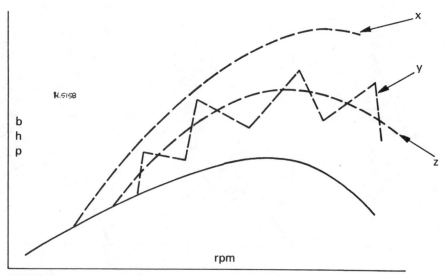

Turbocharger sizing is betrayed by the power curve:
solid line shows standard power curve
dotted lines show variations of sizing
x - good sizing
y - poor sizing, intersurge or stall, turbo likely to destroy itself
z - poor sizing, off islands of efficiency therefore adiabatic efficiency may be 20 – 30% lower
  detonation likely to occur

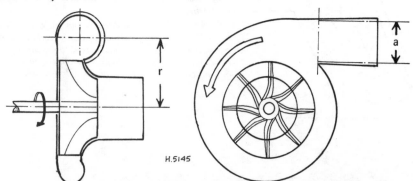

The A:R (area:radius) ratio is crucial to turbine housing choice

sizes in a mechanical supercharger drive. It is a surprisingly straightforward job, one that can be undertaken in the course of road or track trials. Given some practice and a pair of asbestos gloves, the job can be completed in a few minutes.

The object of the exercise is to contrive increased brake mean effective pressure (and hence torque and power) at whatever engine speeds it is most desired. The tendency in the past has always been to look for increased power at maximum-power engine speed — perhaps understandably in racing applications, but with deplorable results in roadgoing cars. It is a forgivable approach perhaps because it is fairly easy: the gradual decline in the gradient of an unsupercharged engine's power curve as the speed rises is due to the engine's increasing difficulty in breathing through a necessarily constricted inlet system, and it is

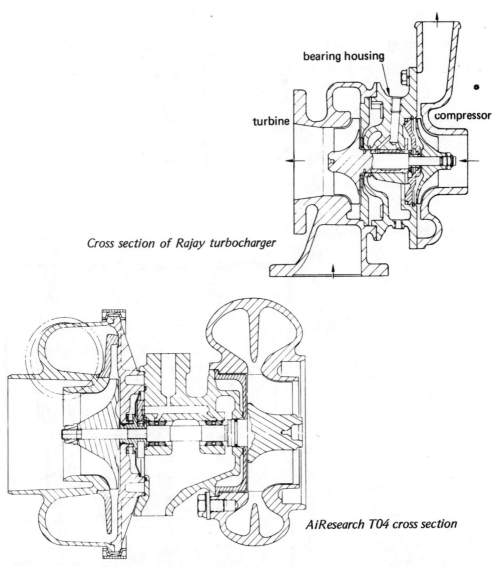

bearing housing

turbine

compressor

Cross section of Rajay turbocharger

AiResearch T04 cross section

the simplest thing in the world for forced induction to restore the sagging volumetric efficiency at these high speeds. To gain more low-speed torque, boost at those speeds should be increased by adopting a turbine housing featuring a lower ratio of area to radius; but then there is the danger that, because of the sky-rocketing delivery characteristics of the speed-sensitive centrifugal impeller, the boost pressure will climb to dangerous levels at higher engine speeds. In this case the flow capacity of the system must be limited by some form of strangulation, the simplest being a restrictive carburettor choke size, an exhaust outlet constriction or even a baffled silencer.

With this kind of matching, engine response to throttle opening is usually superior to that offered by a matching in which boost is arranged for top-end power. It is no substitute

87

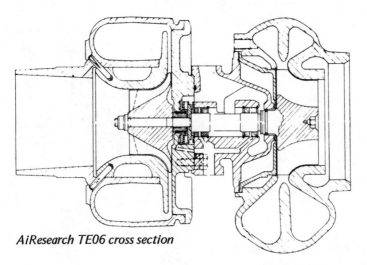

AiResearch TE06 cross section

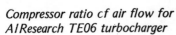

Compressor ratio of air flow for
AiResearch TE06 turbocharger

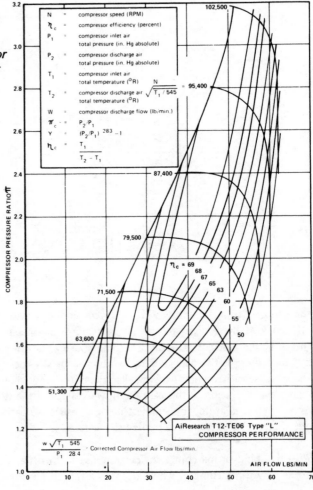

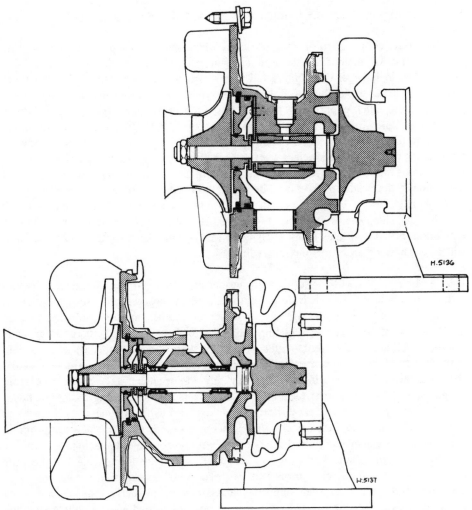

H.5136

H.5137

Typical examples of the Holset 3LD and 4LE turbochargers

The compressor wheel and turbine wheel are fixed to opposite ends of a common shaft which rotates in a central bearing. The complete rotating assembly including thrust and sealing arrangements is known as the rotor assembly and is designed to rotate at very high speeds. (The 3LD rotor revolves at speeds up to 120,000rpm under normal conditions.) Welded together to form a single part, the shaft and turbine wheel is dynamically balanced as a combined unit. The compressor wheel is made as a separate component and is also balanced dynamically. A new compressor wheel can,therefore, be fitted to any new turbine wheel and shaft assembly without special balancing equipment although it is advisable to check balance rotors after long service or possible damage. Also mounted on the shaft is a thrust ring and grooved sleeve which accommodates a sealing ring at the compressor end. A piston ring type oil seal is provided at the turbine end.

The bearings are fully floating and the stability of the rotor assembly is maintained throughout its speed range by the oil films formed between the bearing, the shaft and the housing. Stabilising forces are generated upon the establishment of oil pressure and commencement of rotation; while the unit is stationary, however, a certain amount of play can be felt in the rotor, which is normal.

Each turbocharger manufactured by Holset is individually tested on specially designed equipment. In view of the high rotational speeds the turbocharger housing is designed to retain a burst rotor, although the design of the bearing system would normally prevent the rotor reaching burst speed in service. There is, therefore, an adequate safety margin.

89

for the correct choice of rotor sizes, however, which should be kept as small as possible in the interests of minimal inertia, even if this means having two turbochargers in parallel instead of a single larger one. Duplicated turbochargers are often desirable in any case to simplify the exhaust plumbing, for the piping should not be longer than necessary. Long pipes offer too much opportunity for heat loss, for a turbocharger relies on high turbine inlet temperature: we want the exhaust gases to expand through the turbine, not in the exhaust manifold. For the same reason it is desirable to lag the exhaust pipes, asbestos string being perhaps more efficient than any other method and uglier than most. It ought to be obvious that the exhaust system is no place for the turbulent flow sought in the inlet manifolding: the header pipes should be as 'mellifluous' as possible. There is no need for tuned lengths: resonance is just as undesirable on the exhaust side of the cylinders as it is on the inlet side, and it is better for the exhaust system to be designed for kinetic extraction with the header pipes discharging into a collector through which each blast will induce better flow of the one succeeding it.

If it can be arranged without too lengthy header pipes, and without dictating too large a turbine to handle the flow, it is desirable to lead all the headers into a single confluence upstream of the turbine inlet. If, for one reason or another, multiple turbochargers are used, care must be taken in grouping the pipes to secure equal discharge intervals if possible. In the unlikely event of this being done with a four-cylinder engine, it would mean that the first and fourth cylinders should exhaust into one turbine, the second and third into the other. Two turbines for a six-cylinder in-line engine are easy, the pipes from cylinders one, two and three feeding one, and the remainder feeding the other; but if three turbines are to be fed, they should be linked with cylinders one and six, two and five, and three and four respectively. V8 engines may or may not demand a compromise: those with a single-plane crankshaft, almost entirely confined to racing cars, can be treated as though each bank of cylinders were a separate in-line-four engine, but the more common V8 crankshaft, with throws in two planes 90° apart, strictly calls for cross-coupling of the pipes between opposite banks if equal discharge intervals are to be arranged. This makes the pipework altogether too complicated and for that matter too long, and the objection is commonly ignored without running into too much trouble.

Important though the rotational inertia of the turbine and compressor are, they are not the only crucial parts of the system in ensuring rapid throttle response. What are called the 'controls' matter as much as the basic matching, and these can range from a simple restriction, either in the carburettor or the exhaust system, to complex circuitry full of by-pass gates and trip valves.

The crudest and least excusable such controls are simple blow-off valves which can open against pre-set spring pressure to limit either boost or exhaust pressure. Venting unburned or burned charge direct to atmosphere is at least anti-social, and usually also inefficient, if only because of the danger of the valve going into resonance with some unaccounted harmonic as described in a previous chapter. Elaborated versions of the same thing can work quite well, however, the extreme case being exemplified by the Broadspeed circuitry

Turbine and waste gate exhaust pipes are nearest the camera in this 1962 Oldsmobile installation

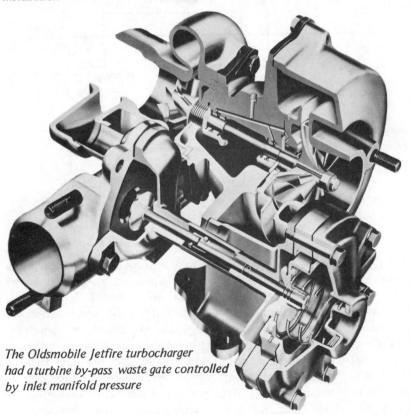

The Oldsmobile Jetfire turbocharger had a turbine by-pass waste gate controlled by inlet manifold pressure

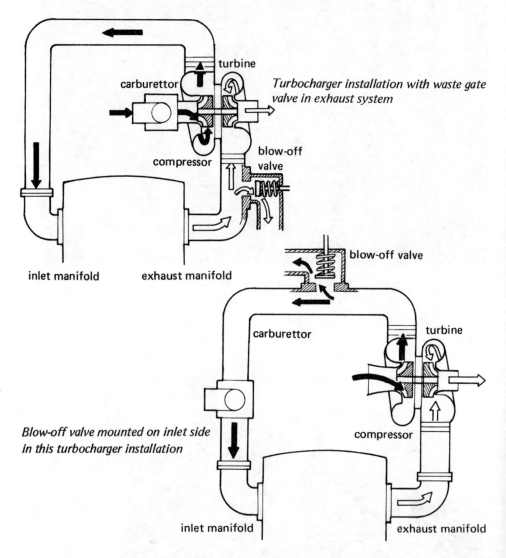

Turbocharger installation with waste gate valve in exhaust system

Blow-off valve mounted on inlet side in this turbocharger installation

hitherto applied on a custom basis to Ford engines by that company and more recently on a production footing for Opel. The turbocharger pressurises a plenum chamber surrounding the standard carburettor, downstream of whose throttle butterfly is a vacuum connection to a valve adapted from an SU carburettor top. When the throttle is closed, the vacuum behind it trips the valve which opens a by-pass through which the air delivered by the compressor is re-cycled to the impeller intake, while the turbine is free to continue spinning under the impulsion of such exhaust gases as are still delivered to the turbine. This re-circulation of the compressor delivery relieves the impeller from contra pressure and concomitant braking effect, perhaps at some cost in reheating the re-cycled air and thus raising the impeller housing temperature; on the other hand, the

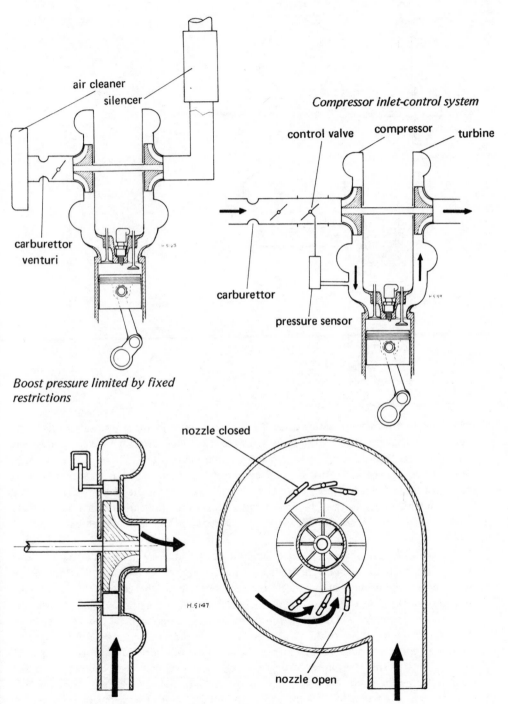

air cleaner

silencer

carburettor
venturi

*Boost pressure limited by fixed
restrictions*

Compressor inlet-control system

control valve   compressor   turbine

carburettor

pressure sensor

nozzle closed

nozzle open

Turbine with variable-area nozzle (vortex throttling)

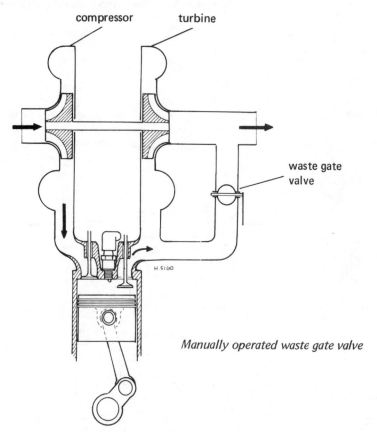

compressor   turbine

waste gate
valve

H.5160

Manually operated waste gate valve

freedom with which the rotating assembly can carry on spinning suggests that this is one
type of installation where a little more than the usual inertia might be welcome. At any
rate, when the throttle is opened again the vacuum behind it is relieved, the valve snaps
shut, and boost pressure is restored to the carburettor plenum chamber. To stop it blowing
the petrol back down the pipe, there is a pressure bleed from the plenum chamber to a
fuel valve which ensures that the fuel delivery pressure to the carburettor float chamber
is always higher by a fixed amount than the air pressure in the plenum chamber.

The whole object of this system was to eliminate the time lag in throttle response that
had always been the bane of turbocharged petrol engines in the past. That harrowing
hiatus between question and answer, between the jab of the accelerator to summon
urgently needed power and the eventual eruption of even more power than you wanted,
two or three seconds too late for it to do any good, is decidedly not a characteristic of
this installation. Response to the throttle is virtually immediate, in time if not in quantity,
for power and pressure continue to rise for a little while after the throttle is opened,
whereas in a normally-aspirated or some mechanically-supercharged engines the brake
mean effective pressure (bmep) is an almost instantaneous function of the throttle
opening. Say what you like about God's atmosphere, it is at least pretty consistent in
pressure!

A form of control that has been much more widely used is the exhaust waste gate, a valve which vents the exhaust system upstream from the turbine and thus by-passes it, either passing the gases to atmosphere or routing them back into the exhaust system downstream from the turbine. This waste gate is not a simple blow-off or pressure relief valve, though it can work in that mode; in more refined systems it is controlled by boost pressure, so as to open when demand is low, relieving the exhaust turbine and preventing it from driving the compressor too fast. An open waste gate reduces back pressure, exhaust valve temperature, and fuel consumption, during cruise conditions; it can also be set to open when flow through the compressor reaches a chosen limit, and thus to give greater boost pressure at moderate engine speeds than at maximum speed, ensuring high back-up torque for roadgoing flexibility. Among numerous users of this system were Porsche, who retained it when eventually they marketed their Turbo version of the 911 road car with a by-pass system that was essentially similar to Ralph Broad's, except that it drew air through a Bosch injection system instead of shoving it into a carburettor. In some cases they found that this compressor by-pass system made their exhaust waste gate redundant: deletion of the gate and its control apparatus saved them quite a lot of expense and complication — and the fact that they are not a firm to be deterred by either suggests that it is better to rely on by-passing the compressor than on by-passing the turbine. To do so with a carburettor on the suction side of the compressor would be difficult, though certainly not impossible; in really high-boost applications where charge cooling is of crucial importance, the recirculatory inlet system might be less attractive.

Using the charge to cool the engine, as opposed to relying on the fuel to cool the intake air, is incidentally not so easy to arrange on a turbocharged engine as in a mechanically supercharged one. The exploitation of a long blow-down during the overlap of inlet and exhaust valve opening might reduce the temperature of critical areas in the combustion chamber, but would also lower the exhaust temperature to the detriment of turbine performance. It is often necessary for a turbocharged engine to have a rather mild camshaft with no more overlap than a gentle touring car might boast, and with a fairly short exhaust valve opening; but since this means that valve acceleration will have to be greater, if the lift and reseating is to be completed in less time, it follows that the turbocharged engine may be more limited by its valve gear than the mechanically supercharged, or for that matter the unsupercharged, engine.

Camshaft design may itself be considered as a form of control, though not an attractive one. Others that remain are the expensive and mechanically unreliable variable-area nozzle for the turbine, working similarly to the vortex throttling device on the centrifugal supercharger described in Chapter 6. With this control, all the gas goes through the turbine at all times and none is by-passed, only the speed of the turbine being controlled by the variations in nozzle area as the vanes are altered in pitch. It is a very good idea that has yet to be effectively carried out on an exhaust turbine; meanwhile the same sensors and servos that control it in relation either to turbine speed, pressure ratio, or air flow, can work a waste gate equally well.

95

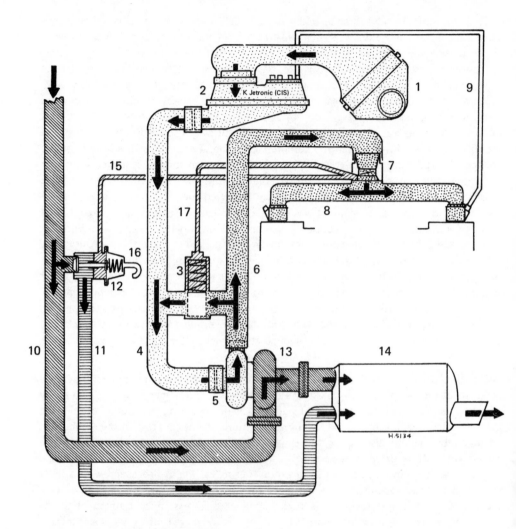

Controls in the Porsche 911 Turbo

The engine draws in atmospheric air through air filter (1), mixture control (2), and induction pipe (4) which then flows through compressor (5) of the supercharger, pressure line (6), throttle housing (7), air manifold (8) and enters the engine.

The engine exhaust gases pass through exhaust manifold (10), supercharger turbine(13), muffler (14) and are then discharged to atmosphere. The exhaust gas flow drives turbine (13) which again drives compressor (5), supplying compressed air to the engine. The supply pressure of compressor (5) is limited by by-pass valve (12) in exhaust manifold (10); when the supply pressure of compressor (5) exceeds a predetermined value, by-pass valve (12) is opened by the excess pressure in control pipe (15) so that the exhaust gas flow passes through by-pass line (11) around turbine (13) directly to muffler (14). For maintaining the supercharger speec, eg under coasting conditions or to ensure quick engine response when accelerating, a connection pipe with blow-off valve (3) is provided between induction pipe (4) and pressure line (6). With the throttle in closed position, blow-off valve (3) in control line (17) is opened due to the differential pressure so that the inlet air passing around compressor (5) ensures the required supercharger speed

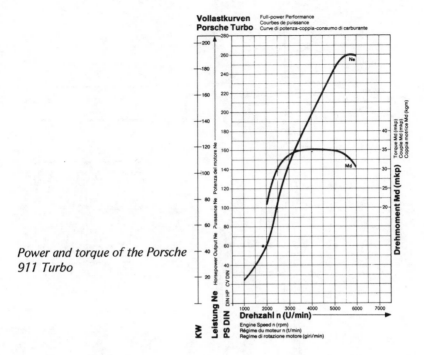

Full-power Performance
Courbes de puissance
Curve di potenza-coppia-consumo di carburante

Ne

Md

Drehmoment Md (mkp)

Torque Md (mkp)
Couple Md (mkp)
Coppia motrice Md (kgm)

Drehzahl n (U/min)

Engine Speed n (rpm)
Régime du moteur n (t/min)
Regime di rotazione motore (giri/min)

Leistung Ne

Horsepower Output Ne   Puissance Ne   Potenza del motore ne

KW   **PS DIN**   DIN HP   CV DIN

Power and torque of the Porsche 911 Turbo

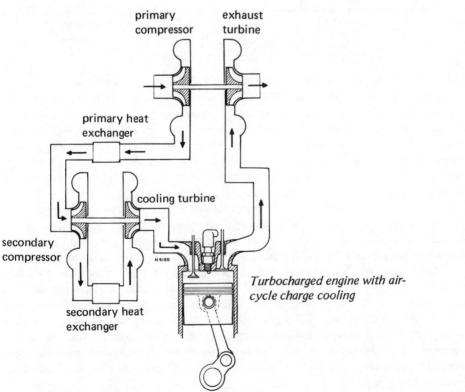

primary compressor

exhaust turbine

primary heat exchanger

cooling turbine

secondary compressor

H.5150

secondary heat exchanger

Turbocharged engine with air-cycle charge cooling

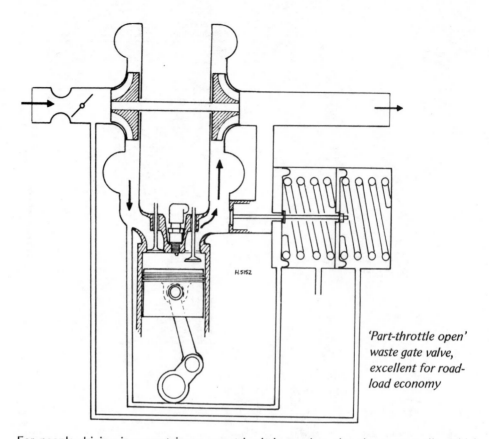

*'Part-throttle open'
waste gate valve,
excellent for road-
load economy*

For people driving in mountainous countries it is worth noting that a controller which senses boost pressure ratio to control turbine speed will work effectively regardless of altitude, but is affected by changes in ambient air temperature that commonly accompany changes in altitude, anyway. For a vehicle subjected to extreme altitude changes the best device might be one that responds to absolute boost pressure; it is a simple aneroid which compares boost pressure with a vacuum and is not affected by changes in ambient pressure. Where no such control is exercised, there is some danger of the turbine overspeeding at high altitude, due to the reduced atmospheric pressure increasing the pressure difference across the turbine. With an aneroid control it is possible to secure sea-level power at high altitudes (as is done in aircraft) without excessive boost at low altitudes. In conditions that are unlikely to change materially, as in fairly flat country and reasonably temperate climates, a simpler form of control is to relate to boost pressure ratio rather than absolute pressure. Indeed for those applications (more numerous than most turbocharging enthusiasts will admit) where greater boost is desired at medium engine speeds than at higher speeds, to provide torque back-up for acceleration and easy vehicle control, the turbine speed can be regulated by a waste-gate set to open when a sensor indicates a high mass flow through the inlet manifold, which can be conveniently measured a little downstream from the compressor discharge. This kind of measurement is in fact just what is done by the Bosch fuel injection apparatus used *inter alia* in the Porsche 911 Turbo.

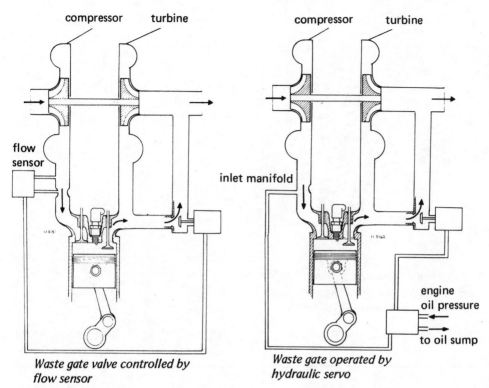

compressor turbine

flow sensor

*Waste gate valve controlled by
flow sensor*

compressor turbine

inlet manifold

engine
oil pressure

to oil sump

*Waste gate operated by
hydraulic servo*

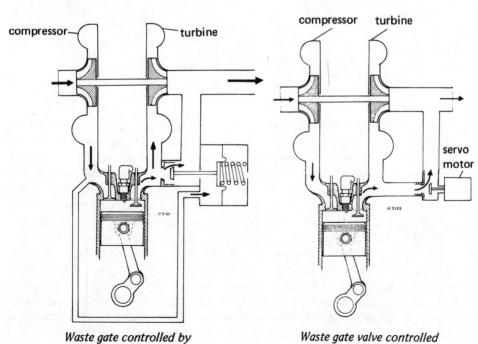

compressor turbine

*Waste gate controlled by
inlet manifold pressure*

compressor turbine

servo
motor

*Waste gate valve controlled
by servo motor*

99

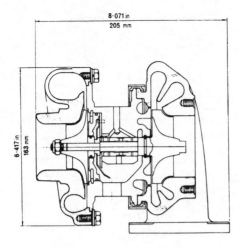

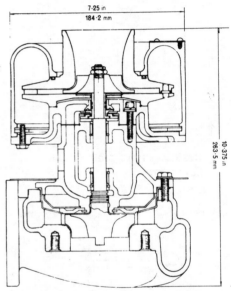

*Modern Holset turbocharger with 3 inch rotors
(top) is much simpler and more compact
than its equivalent of 1957 (bottom)*

As was pointed out earlier, the waste-gate is a valuable adjunct in a turbocharging installation for a road car, in which duty the waste-gate should be open most of the time except during acceleration. Perhaps the nicest contrivance to achieve this is the part-throttle-open valve which allows the waste gate to open whenever the pressure drop through the carburettor is greater than a stipulated amount. This is measured by simply comparing pressure at the engine inlet port (or anywhere conveniently downstream of the compressor) with pressure just downstream of the suction carburettor's throttle slide or butterfly. These two pressures are piped to the outermost of three chambers, separated by spring-loaded diaphragms, the middle chamber being vented to atmosphere. More than a certain amount of depression behind the throttle, or more than a certain boost pressure downstream from the compressor, will override one or other of the springs and lift the stem of the poppet valve that functions as a waste-gate for the turbine by-pass. It is an arrangement that gives good low-speed performance without danger of over-boosting at high speed, and a cruising fuel consumption that is comparable with, and may even be better than, that of the same engine naturally aspirated.

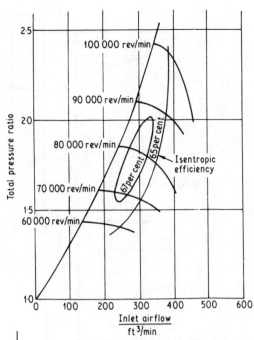

*Characteristics of 3 inch Holset
compressors have likewise improved
enormously between 1954 and 1971*

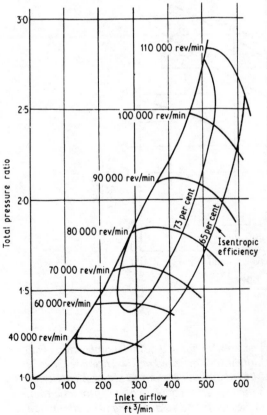

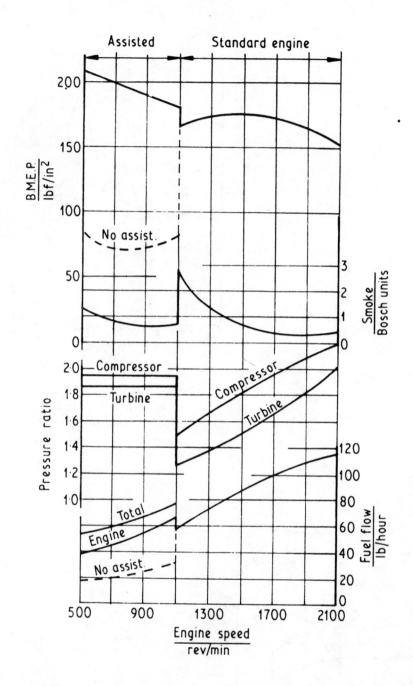

A turbocharger is just a small gas turbine, so why not add a small combustion chamber (in parallel with the engine) and so maintain turbine speed at low engine speeds? The effect, when tried by Holset on a 12¼-litre diesel engine, was to produce an astonishing torque curve like that derived through the converter coupling of an automatic transmission. At the same time, problems of lag, surge and cold starting were eliminated

Chapter 8
Compounding
and differentiating

No matter how painstaking the sizing and matching, how thorough the attention to manifolding and carburation, simply to tack a supercharger or a turbocharger onto an engine is not going to get the best out of it. The device is a pump that can in a sense be considered a kind of engine in its own right: and it is reasonable that when coupling together two engines of different kinds, the connections must be both supple and subtle if the resulting complex variable is to be efficient in all modes of operation. How much improvement can be wrought at the expense of a little complication has already been shown in Chapter 2 where two-stage supercharging was discussed; and in Chapter 6 we saw how, for some purposes, two stages of compression, one by mechanically driven blower and one by turbocharger, were necessary in combination. This two-stage turbo-supercharging was to make the most of the two-stroke engine, but it has been demonstrated in practice that if both types of compressor drive are employed, it is nevertheless not strictly necessary to have them drive two compressors.

The engine that best demonstrates this is one that is an oddity in numerous other respects as well. It is the Mitsubishi 10 ZF type 21 WT tank engine, an air-cooled two-stroke V10 diesel that develops 858 horsepower from 21.5 litres and is extremely compact. The alimentary peculiarity of the engine is that there are two exhaust turbo-chargers at the front end of the engine, driven mechanically through a torque-limiting clutch by the crankshaft. The object of this was to get rid of the positive-displacement type of scavenge blower conventionally thought necessary for the high-speed two-stroke engine, and to keep the supercharging system as compact as possible because space is

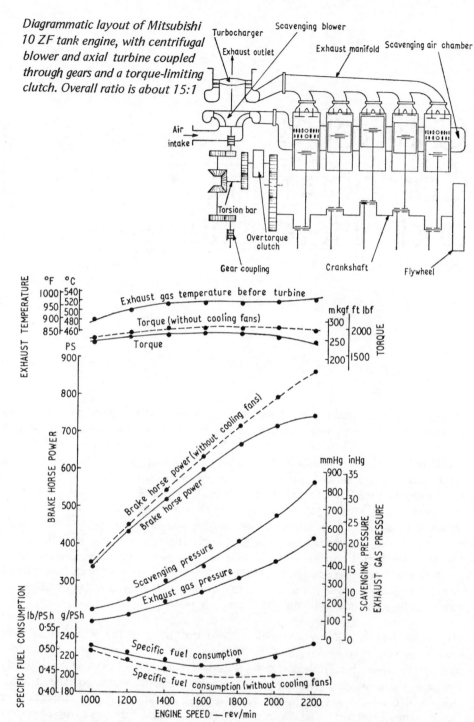

Diagrammatic layout of Mitsubishi 10 ZF tank engine, with centrifugal blower and axial turbine coupled through gears and a torque-limiting clutch. Overall ratio is about 15:1

Mitsubishi 10 ZF engine performance: note that cooling fans impair the specific fuel consumption as they absorb power

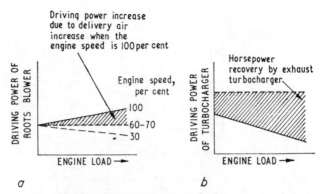

Comparison of supercharging systems studied for 10 ZF Mitsubishi

a   Supercharging system with Roots blower in series
b   Mechanically driven turbocharger system

always at a premium in tanks. An interesting feature of the turbochargers is that while the compressor is centrifugal, the turbine is an axial-flow type — a feature that also distinguishes another exceptionally efficient and unusual high-performance diesel, the Napier Deltic. In the Napier, clever use was made of torsion bars in the blower drive, and in later models of axially flexible helical gearing, to protect the high-speed turbine and impeller from shock loads; the same applies to the torque-limiting clutch in the Mitsubishi which is there simply to filter out overloads, not to modulate the performance.

It is interesting to compare the mechanically driven turbocharger with the more conventional arrangement of an exhaust turbocharger and a mechanically driven Roots blower in series. With the latter arrangement, in the higher speed range, the rotating speed of the turbocharger is increased as the engine load increases, the result being higher air pressure at the delivery from the centrifugal impeller, and hence a higher pressure at the inlet of the Roots blower. It follows that at a higher engine speed the power necessary for driving the Roots blower would increase as the engine load increases. With a mechanically driven turbocharger, on the other hand, the exhaust gas temperature and pressure are increased as the engine load increases: thus the turbine draws more power from the exhaust gas, with the result that the power required to drive the blower is reduced. It is worth pointing out that if these two systems are designed to give the same scavenging air pressure at full load (which may not necessarily be desirable) the mechanically driven turbocharger system will be required to consume more power in the part-load condition compared with the conventional arrangement.

The Mitsubishi system is not without its shortcomings. Determining the gear ratio ·between crankshaft and turbine shaft has more than its fair share of headaches: raising   **105**

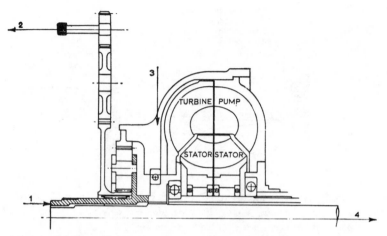

Combined differential and torque converter arrangement

the ratio to give a higher turbine speed might well produce more scavenging air and a lower exhaust temperature, which in the case of the diesel brings the opportunity to increase maximum power output by increasing the quantity of fuel injected, limited only by peak exhaust gas temperature. However, the maximum explosion pressure in the cylinder would be raised, specific fuel consumption would be poor, and the part-load power consumption of the blower would be even higher.

However, there is great attraction in coupling the turbine to the engine crankshaft for, as was pointed out in Chapter 7, the energy available at the turbine is frequently much greater than that which is necessary to drive it. This is one of those situations that begs to be exploited by enthusiastic development, the logical consequence of which is to produce something quite different in the end from what we set out to achieve at the beginning. Uninhibited development of the turbocharger ought in fact to lead to the eventual elimination of the piston engine!

The progression seems perfectly logical, leading at an early stage to the compound engine. Consider what happens when a turbocharged engine is uprated: gradually you build up the mass flow in the exhaust system and perhaps also the back-pressure, thus contriving to make the exhaust turbine more and more powerful so that it can drive the supercharger impeller to produce more and more boost pressure. Carry on like this and you ultimately reach the stage where the exhaust turbine is developing as much power as the engine itself. At this stage it becomes logical to transpose the duties of the turbine shaft and the crankshaft, arranging for the turbine shaft to provide propulsive power for the vehicle, and for the crankshaft to drive the supercharger.

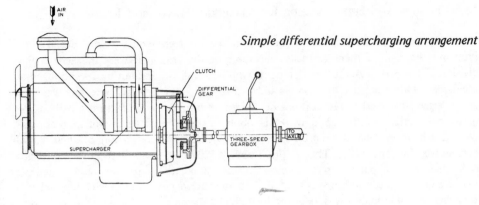

Simple differential supercharging arrangement

CLUTCH

DIFFERENTIAL
GEAR

THREE-SPEED
GEARBOX

TO
AXLE

SUPERCHARGER

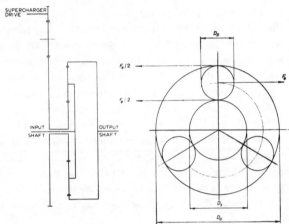

Gear arrangement for a differential diesel

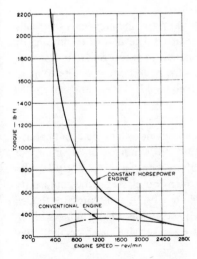

Contrasting torque curves of conventional and constant-horsepower engines reveal the desirability of substituting a differential compound turbocharging concept for the time-worn conventional multi-ratio transmission

107

Turbocharging and Supercharging for Maximum Power and Torque

Inevitably it was those efficiency experts in the aviation industry who did it first. Probably the first of them to realise the force of the argument were the aero engine division of the Bristol Aeroplane Company, who had it all worked out by 1938 but found themselves with more urgent business and had to shelve the project. After the war they leap-frogged the next stages of engine design, but other manufacturers had in the meantime carried on the good work. Undoubtedly the most successful of these was Wright, which evolved the Turbo Compound engine as a conversion of their 2,800 horsepower Cyclone radial. That engine weighed 3,029 lb, giving it a power-to-weight ratio that was not particularly impressive by the standards of its time, and something better and more economical was wanted. As a precautionary measure the cooling fin area of the heads and cylinders was increased by 40% and the crankshaft stiffened to cope with the anticipated extra power; then the cylinders were divided into three groups of six and the exhaust gases from them were piped from each group to one of three turbines. Each turbine was connected to the engine's crankshaft through fluid couplings which were fed with engine oil — it will be recalled from Chapter 6 that the simple fluid coupling, whatever the disparity between the input and output shaft speeds, transmits exactly the torque delivered to it, and is therefore an ideal means of communication for this particular task. The engine crankshaft itself drove a supercharger which, as in the case of Mitsubishi, took its share of the power developed by the exhaust turbines and fed back to the crankshaft. It made a tremendous difference, as of course it should, for a lot of power is usually wasted down the exhaust pipes, even in a conventionally turbocharged engine. In the case of the Wright Cyclone, this feed-back raised the maximum power to a good 3,500 horsepower, and the thermal efficiency of the engine took a smart turn for the better as the specific fuel consumption dropped to a frugal 0.38 lb per horsepower hour.

To the aviation trade this was a marvellous shot in the arm, for the engine was virtually no bigger than the standard Cyclone and could easily replace it. The aircraft could then fly either 20% further on the same fuel, or cover the same distance on proportionately less fuel, either way spelling out good business. In motoring terms the equivalent would be to make the car capable of reaching a speed 9% higher, or of travelling 20% further on a given fuel load or of being able to carry 17% less fuel.

For a super-efficient car one would hardly take the Wright engine as a model: commercially viable it may have been, but in engineering terms it was little more than a lash-up. For a real virtuoso performance one must look, as has so often been the case, to Napier, who produced in the Nomad a compound turbocharged aero-engine of staggering complexity and highly creditable refinement. Everything they did in the 1930s and thereafter tended to be like that, not least their 5,500 bhp Sabre engine; and it was when toying with an abortive two-stroke version of that incredibly effective sleeve-valve machine that they found their clues for the compound turbocharged one.

Napier reasoned that an exhaust turbine would inevitably create back-pressure in the exhaust system, and went on to argue that if their sleeve valves were not in the least put

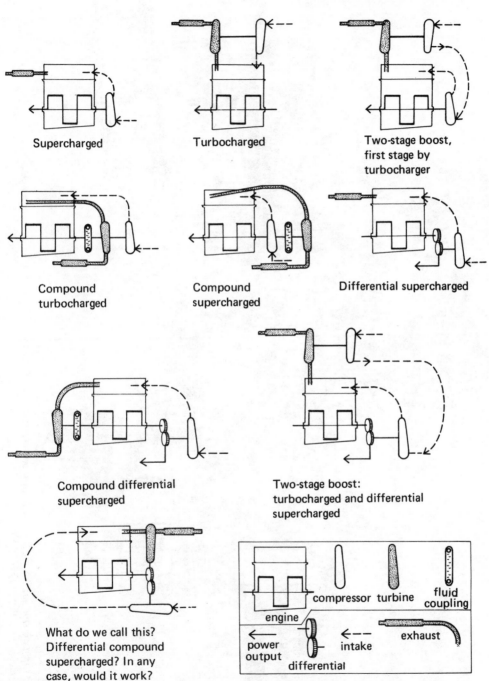

Supercharged

Turbocharged

Two-stage boost,
first stage by
turbocharger

Compound
turbocharged

Compound
supercharged

Differential supercharged

Compound differential
supercharged

Two-stage boost:
turbocharged and differential
supercharged

What do we call this?
Differential compound
supercharged? In any
case, would it work?

compressor turbine fluid coupling

engine

power
output

intake

exhaust

differential

*How cost-effective can a complex system be? And how eventually costly and inept the
simple alternatives are!*

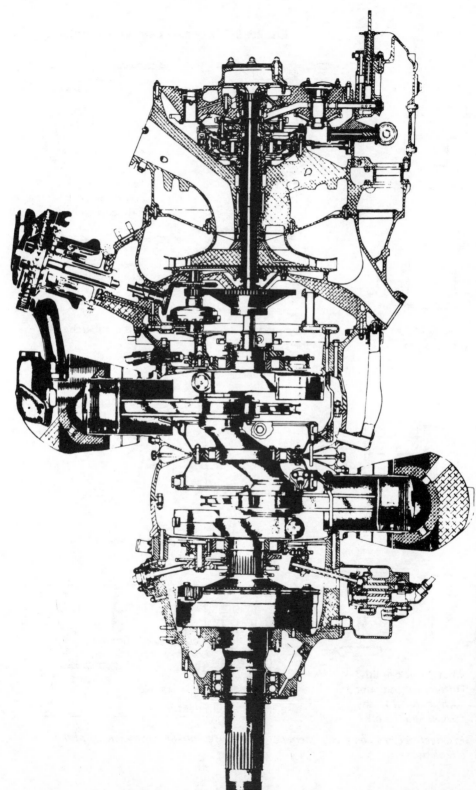

Section through the Wright Turbo Compound aero engine, which so dramatically improved long-range air-craft performance in the late 1940s

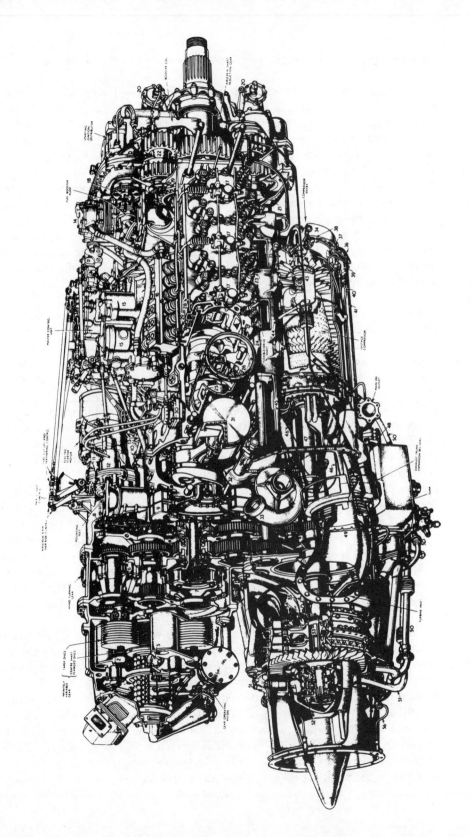

Perhaps the ultimate in complexity, efficiency, and cost, the Napier Nomad failed commercially despite espousing the same principles as the Wright Turbo Compound

out by back-pressure, a two-stroke engine would positively revel in it. Further, they reckoned that for high thermal efficiency and large mass flow (important with turbines) a diesel engine was the thing, so they made it a two-stroke diesel. By the time the designers had finished their syllogisms, they had made the Nomad a 12-cylinder horizontally-opposed two-stroke diesel supercharged to colossal pressures, as much as eight atmospheres. The exhaust gases from the 12 cylinders were piped to a three-stage axial-flow turbine which had, mounted on a concentric shaft, a 12-stage axial-flow compressor to feed more air into the engine. To make sure that nothing was wasted this turbine shaft was linked to the engine crankshaft by a fluid coupling — and to make doubly sure that nothing was wasted, the last remaining bit of energy in the exhaust gases after negotiating the turbine was exploited in the form of jet thrust, which amounted to about 250 lb at maximum power. In this form the engine pushed out 3,570 bhp — which must have been intensely irritating to the tidy-minded, for it weighed 3,580 lb!

The beauty of a situation like that is that a brief but massive investment in compound interest provides the necessary bonus. By this time Napier were enormously interested in compounding and went after a really big further increment in power (the extra 10 bhp was neither here nor there) in the most adventurous way imaginable. Most of their Nomad engine remained the same, but an intercooler was added to lower the temperature of the charge between compressor and cylinders, and then they added re-heat to the exhaust manifold by the simple process of injecting extra fuel there and burning it. This was something they could do most effectively because the engine was a diesel, and since a diesel cannot normally burn more than about 70% of the air it breathes, there was plenty of oxygen still available to support the combustion of the extra fuel. The extra engineering involved little weight penalty, only 170 lb, but the payoff was fantastic — another 930 bhp. Thus the specific weight came down to 0.83 lb per horsepower, surely the best diesel figure ever. As for the bmep, it worked out at no less than 345 lb/in^2, a colossal figure for a two-stroke — a corresponding four-stroke would have to operate at 690 to produce the same results. This is a figure that deserves to be compared with something, but it is difficult to decide what. A no-holds-barred dragster guzzling a typical mixture of nitro-methane and boot polish might sustain such pressures for several seconds; a Formula 1 Grand Prix car might average 200 lb/in^2 for the greater part of two hours; the V16 BRM managed a highly respectable 460, but it was a notable failure.

There is no denying the fact that the Napier Nomad was a failure too. By the time it was built and running — and by all accounts it was a pig to start — the British aircraft industry had already espoused the gas turbine, and people in high places did not want to know piston engines any more. They were quite right, of course, and if Napier had pursued their theory further they would have been forced to agree — as perhaps they secretly did, since they were hedging their bets with the Eland gas turbine. The point is that when you start using your exhaust turbine as a power source in this way, the piston engine ceases to be a torque developer and becomes merely a gas producer, a clumsy and mechanically horrendous ancillary to the smoothly spinning turbine. If you must have a gas producer, why not get rid of all those nasty reciprocations and rely on turbines to do

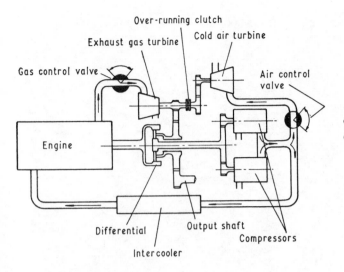

An important feature of the compound differential is the auxiliary cold air turbine

Characteristics of a mechanically driven turbocharger, when engine speed is constant, are greatly influenced by impeller and diffuser design. Blower B is better (because it provides more back-up torque) than A in diagram c

a	Flow characteristics in turbine
b	Flow characteristics in cylinder
c	Scavenging blower characteristics
G_s	Mass of scavenging air
G_t	Mass of exhaust gas
P_t	Exhaust gas pressure absolute
P_o	Ambient pressure
T_t	Absolute temperature of exhaust gas

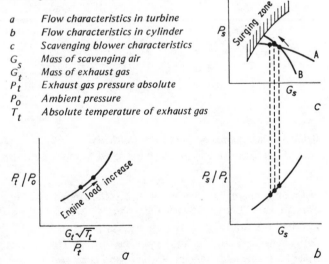

all the compression instead of just some of it? Why not, indeed? — you are doing no more than re-inventing the gas turbine, a step which Bristol had foreseen in 1938 and which I had in mind when, earlier in this chapter, I suggested that development of the turbocharger would mean the end of the piston engine.

This is not to say that even more complex combinations of piston engine and turbo-superchargers cannot be devised. It has been done at research level in the form of the differential compound engine, a concept dating from patents by Invernizzi in 1925 and Weber in 1927. More work on the idea was done in the 1950s, in Sweden and especially

113

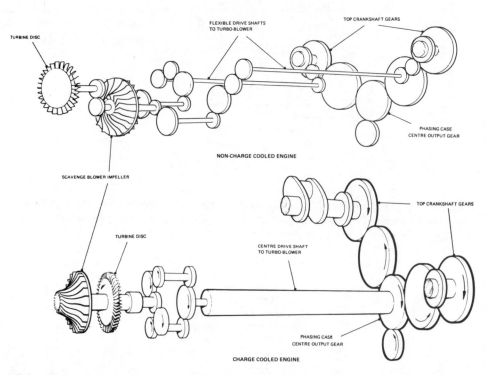

While adding charge cooling to the Deltic engine, Napier revised the drive to the compound turbocharger to solve torsional vibration problems

in France, where Berliet were probably the first to make it work in a vehicle. The leading protagonist at Berliet was Dr. Ing. Glamann, who was later consulted by Perkins after the idea had been developed in a theoretical paper in 1963 by Professor Wallace, working at Bath · University of Technology. The differential compound engine is one in which three basic elements (engine, turbine, and compressor) are connected through differential gearing. The exhaust gases from the engine are fed to a turbine, the output shaft of which is geared to the output shaft of the engine, and to the planetary carrier of an epicyclic differential gear in which the annulus is on the engine output shaft and the sun wheel drives the supercharger.

The idea was to eliminate the need for multi-ratio gearboxes or their hydrostatic or hydro kinetic equivalents, and initial tests broadly confirmed the predictions of torque: speed and power: speed relationships which were expected to be appropriate to traction-type prime movers. The feasibility of the idea was publicly demonstrated in

1964 by the Perkins company in a truck powered by what they called simply the "differential diesel". The arrangement used by Professor Wallace secured particularly high torque at low output shaft speeds (the Perkins proved to need a two-speed gearbox for fully acceptable performance, and to put it in perspective a conventionally engined truck of similar power and weight needed at least a six-speed gearbox to be satisfactory) by the incorporation of an auxiliary cold air turbine which was fed by excess air from the compressors and which in turn fed its output torque into the output shaft of the exhaust gas turbine by way of an overrunning clutch.

Interim test results were impressive, though power and torque fell short of predictions, particularly at high output shaft speed and high boost pressure ratios. Much of this was ascribed to severe parasitic losses, many of which would be eliminated in production, and it was forecast that the incorporation of variable nozzle geometry for the power turbine, not to mention interburning between the compressor and auxiliary turbine, would materially improve the performance of the engine.

There seems no end to the complexities that can be introduced. The last may yet turn out to be one of the most significant, a variable compression ratio. The virtue of this is that it controls peak cycle pressure in such a way as to make it practical to run at bmep values double those of otherwise conventional engines of similar size and weight. The mechanical advantages are not inconsiderable: heavy loads are intercepted before they can be transmitted to the bearings, so their lubricating oil film wedges remain intact, and a similar immunity from transient overloads applies to all other engine components, ensuring long life. In a forced-induction context, which is how the system was developed by Professor Timoney at University College, Dublin, the supreme blessing of a variable compression ratio system that works is that it provides automatic compensation against detonation dangers at high boost pressures, and restores high geometrical compression ratio and corresponding thermal efficiency when the engine is running at part load.

To achieve this, Professor Timoney had recourse to that standby of the academic engine-research man, the three-cylinder six-piston Rootes compression-ignition two-stroke. Its most unusual feature is the use of rocker arms connecting the piston connecting rods to larger and more conventional connecting rods swinging about the pins of a conventional crankshaft located below the block of horizontal cylinders. This arrangement has numerous practical virtues, as well as others that make the layout lend itself to experimentation in many directions: it was the work-horse of Professor Wallace in the development of the compound differential diesel, for example. Professor Timoney, Head of Thermodynamics at University College, Dublin, wanted to make a diesel of high specific output, compact dimensions, and exemplary durability; and the crux of his work lay in the simple exploitation of the Rootes rocker arms to give rapid and precise control of a variable compression ratio.

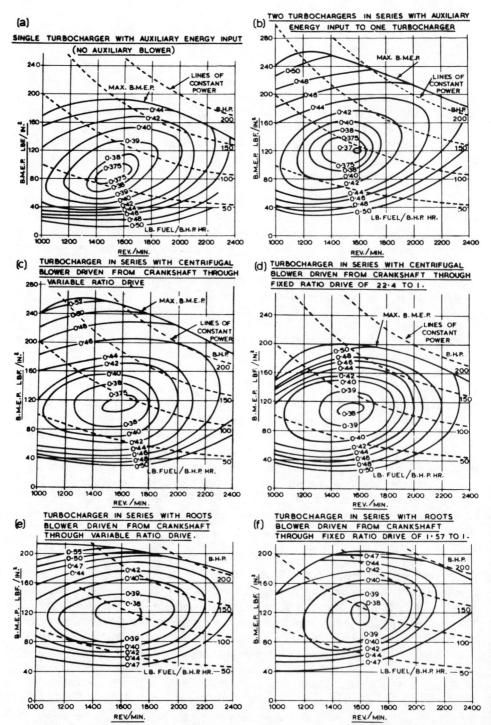

Marked differences in the performance of the Timoney vcr engine accompany changes in blower types and drives

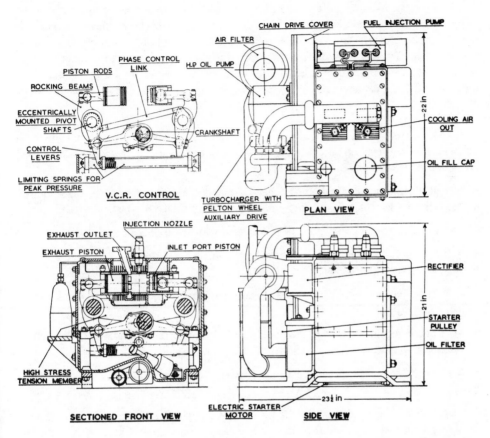

Layout of the Timoney variable compression ratio engine

Two inter-dependent principles are involved. One is the pursuit of high specific output through the combination of a two-stroke cycle and turbocharging, the other is the automatic variation of compression ratio to ensure long service life by controlling loads on the pistons. The mechanical means of achieving this end, sought with so much complication in other designs, are very simple in the adapted Rootes engine, relying on eccentric bushes for the pivots of the rocker arms. The change in compression ratio needs to be responsive to changes in engine stress, which are equivalent to changes in the peak cycle pressure. This makes it possible for the control mechanism to be made automatic: the eccentrics are rotated by levers which, if left unrestrained, would tend to turn under the load applied to the pistons by compression pressure. A spring link between them provides the necessary restraint, and responds directly to the cylinder pressure, adjustment of the spring rate controlling the amount of change of the compression ratio.

117

Turbocharging and Supercharging for Maximum Power and Torque

The response is amazingly fast, faster than could be required of an engine in service. Artificially rapid loading and unloading of the engine on the test bench proved that the compression ratio could be raised from 9.5 to 16.5:1 in six cycles of the crankshaft, and reduced again in eight. As a consequence it proved impossible to produce audible diesel knock in the Timoney engine. The interesting and perhaps unexpected corollary of this is that the load on the eccentric shaft control levers, and hence on the spring, must be a function of the peak cylinder pressure and not of the mean pressure, as might be expected.

With this valuable refinement added to all the others, there is no telling what might be achieved with turbocharging, especially of the compound variety. In fact Timoney tried half a dozen different induction systems:

1. A single turbocharger with auxiliary energy fed to it through an oil turbine incorporated on the turbocharger shaft from a high-pressure oil pump driven by the crankshaft.

2. Two turbochargers in series, one of them fitted with an auxiliary engine system

3. A turbocharger in series with a centrifugal supercharger driven from the cra through an infinitely-variable-ratio drive.

4. A turbocharger in series with a centrifugal supercharger driven from the crankshaft at a fixed speed ratio.

5. A turbocharger in series with a Roots blower driven from the crankshaft through an infinitely-variable-ratio drive.

6. A turbocharger in series with a Roots blower driven from the crankshaft at a fixed ratio.

Examination of the results shows that system No. 2 gave a significant improvement in flexibility and that No. 3 rivalled it in power and economy. It was a fairly elaborate installation, mark you, with an intercooler between the two compressor stages and a further one to cool the charge air before entry to the engine; but that is the way things tend to happen. If one is greedy for performance and unable or unwilling to call on the fuel chemists for aid, to embark on any programme of forced induction is to face a choice of failure or of never-ending complication. Thoreau, who exclaimed "Our lives are being frittered away by detail. Simplify, simplify!" was undoubtedly right; but if and when our cars of the future are blessed with stratified-charge engines, which are inherently immune to knock and to fuel quality and may thus be subjected to any chosen compression ratio or boost pressure within their mechanical abilities to sustain, it may be that forced induction will cease to seem wrong.

Chapter 9
Buyers' guide

Forced induction is something that can be applied to an engine at any stage from the drawing board to the finished vehicle. As will have become clear from the previous chapters, some engines demand careful design *ab initio* for blowing to be effective; others that were never intended to be pressure-charged or scavenged are often amenable to the imposition of a forced induction system as an afterthought. It follows that no thorough list of sources of suitable equipment could be composed without it becoming unwieldy and for the most part superfluous; yet a little guidance is desirable, for superchargers and turbochargers are not things to be bought easily or casually — and if they were, the results would almost certainly betray such carelessness.

For the designer in industry we may assume that appropriate industrial directories are at hand, and that his hand need not be held. For the manager of a small tuning business, or the individual concerned with his own singular (and to him doubtless remarkable) project, to be pointed in the right direction yet without being given a shove is presumably a help. Nor may we overlook the customer who wants to buy a complete conversion, or even a car already fully furnished with blower and ancillaries by its maker.

The last category is the most easily summarised, for today there are no mechanically supercharged cars in production, and only two current 1976 models catalogued as turbocharged. Easily the more outstanding of those two is the Porsche 911 Turbo, which in 3-litre production form (as distinct from the limited series of racing versions) offers not only a lesson in turbocharging for all to study, but also a lesson in vehicle design and manufacture. The man who can afford to order one of these superb machines does not really need a buyer's guide, nor even such recommendations as I can offer from my experience of driving the car in varying conditions and throughout its speed range, which extends to 157 mph; he will know the reputation of the Porsche Turbo, and for him that will be good enough.

The man who would not or could not spend so lavishly upon a car must turn to TVR for the turbocharged choice that is then Hobson's. In 1975 this earnest and durable little firm introduced a turbocharged version of their 3000M sports car, in which the power unit is the Ford 3-litre V6. The turbocharging apparatus is that developed by Broadspeed for application to this engine in any of the numerous cars in which it is original equipment, not only Ford's own but also the Reliant Scimitar GTE. This is a particularly ingenious system, and (as is also true of the Porsche turbocharging layout) has been described at length earlier in this book. The effect on the performance of the Ford V6, and hence on that of the cobby little TVR, can best be described as dramatic; with an increase in power of about 50%, nothing less could be expected.

The Ford V6 is by no means the only current or recent production car engine to have enjoyed the attentions of the turbochargers. The 4-cylinder Fords do too, as have the four-cylinder 1.9 litre engine of the pre-1976 model Opel Manta, and the four-cylinder 2-litre engine of the BMW 2002. The Opel has been produced as a catalogued model by Opel in Germany, where it was equipped with Eberspacher turbocharger; but the German Turbo Manta was not considered satisfactory by the British branch of General Motors, who arranged instead for a Broadspeed installation to be developed for the Manta in its most polished guise, the Berlinetta. This 'Black Manta' (there was no alternative colour) or Turbo Manta Berlinetta, could only be bought as a new car, and then only from selected Opel dealers in Britain; but it was a great success, and about fifty were completed and sold before the basic model was superseded by a new Manta that in turn is intended (at the time this book was completed) to enjoy similar treatment.

In the case of BMW there was never much dissatisfaction expressed over the 2002 Turbo which was again Eberspacher-equipped. However, BMW undertook only a limited batch of these cars before running down the 2000 series to make way for the new 3-series cars that carry similar engines and may in time receive similar modification. In the meantime the firm which inspired both Opel and BMW, Turbo May GmbH + Co KG of Hechingen, continue to provide a specialised service in adapting customers' cars (notably Fords, Opels and BMWs) to turbocharging.

The same is done in Britain by Broadspeed Ltd of Southam, Warwickshire. They are best known for their turbocharging of the Ford V6 in Capris, Granadas and Scimitars, and are a highly responsible firm whose expertise in racing preparation — (they ran saloon racing programmes for British Leyland and Ford of Britain at various times in the past, and are doing it for Leyland again) ensures their competence in raising the

With a maximum speed of 157 mph (253 km/h) the Porsche 911 Turbo is justified in the elaboration of its aerodynamic aids to stability

TVR have successfully produced a turbocharged version of the V6 Essex engined 3000

The engine of the TVR Turbo looks less complex than many others: full throttle acceleration from 400 rpm is available in the high top gear

qualities of a car's running gear, suspension and brakes, to the standards appropriate and necessary for the safe exploitation of the performance yielded by turbocharging.

A newer entry to the lists is Mobelec Ltd of Oxted, Surrey, who already have a fine reputation for their exceptionally good electronic ignition apparatus. They have developed turbocharging kits for the British four-cylinder Fords, employing Rajay turbochargers imported from the USA whereas Broadspeed employ the British-made Holset. Another American instrument, the Airesearch, is favoured by Minnow-Fish Ltd of Lochgilphead, Argyll, whose speciality is the turbocharging of the Chrysler Avenger and whose unique feature is the use of the unconventional but effective Minnow-Fish carburettor.

Finally in Britain there is one firm specialising in supercharging. This is the Allard Motor Company Ltd of Putney, London, who are sole concessionaires in this field for the Shorrock eccentric-vane supercharger. This device is available from Allard to suit car engines ranging in size from about 800 cc to as much as 3 or even 4 litres, and in many cases (including several Ford and British Leyland models) Allard can also supply the necessary manifolding and brackets to complete a ready-tailored installation.

122

Broadspeed turbocharged the Opel Manta Berlinetta, which was recognizable by its special black finish — not to be confused with any red or East Berlinetta

Early version of Turbo May turbocharging in the original Opel Manta

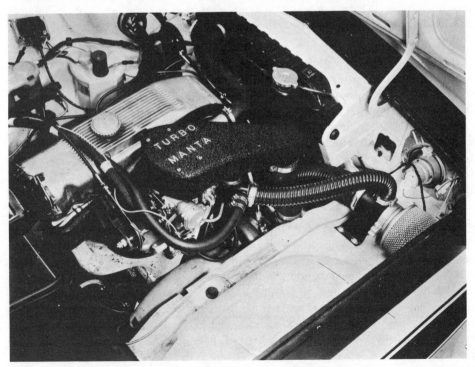

Turbocharging and Supercharging for Maximum Power and Torque

Compared with the wealth of performance shops in the USA this summary may seem lamentably brief, but it must be related to the size of the market. The same applies to the choice of superchargers and turbochargers, which in America is an extensive one, though most of the former are of General Motors provenance (intended for uniflow-scavenged 2-stroke diesels) and most of the latter are either Airesearch or Rajay. Britain has its two-stroke lorries too, however, thus ensuring the continued availability of the big Wade compressor of Roots type. Smaller Roots blowers have in the past been adapted from the Marshall-Nordec aircraft cabin pressurizer, but Broom & Wade superchargers specifically suited to automotive engines have generally proved superior. The BroomWade trade name is now championed by CompAir Industrial Ltd of High Wycombe, Buckinghamshire, who should be particularly noted as manufacturers of a very efficient screw compressor — that is, one of the Lysholm type to which attention was drawn in Chapter 3 — in a considerable range of sizes.

The other British supercharger is the vane-type Shorrock, which as already mentioned is available from the Allard company. Many years' history have led to this device being more free than most vane-types of the traditional shortcomings: highly-loaded rubbing points are eliminated, and the use of ball or roller bearings completes the measures necessary to avoid the lubrication problems that were once the bane of such blowers. There is a tendency for oil to accumulate in the bottom of the rotor housing when at rest, confirmed by a cloud of blue exhaust smoke when starting again; but apart from this brief dosage of an upper cylinder lubricant that may or may not be thought beneficial, the Shorrock is not temperamental.

Turbochargers from the Airesearch range are available in Britain through Garrett Airesearch of Skelmersdale, Lancashire. The importation of Rajay instruments is rather difficult due to the manufacturer's distribution system in the USA. British turbochargers in a wide variety of sizes and types are available from a member of the Lucas group of companies, CAV Ltd of Acton, London, who handle the products of the Holset Engineering Co Ltd of Huddersfield, Yorkshire. Holset rotors come in diameters from 3 to 6 inches, the smaller ones being particularly suitable for car engines and widely favoured by turbocharging specialists of whom Broadspeed have been most conspicuously successful.

After all that has been said earlier in the book about the speeds and loads to which blower rotors are subjected, it should be obvious that any attempt to produce a home-made compressor (let alone an exhaust turbine) is likely to be a failure, and probably a costly one. Nevertheless it should be borne in mind that a serviceable centrifugal supercharger can be built from the bearing housing and compressor section of a turbocharger, remembering that the bearings are not designed to accept heavy asymmetrical radial loadings nor end thrust. As with a normal turbocharger installation, care must also be taken with lubrication provisions: if oil is not present at adequate pressure and flow rate within three seconds of starting, harm is likely to ensue.

That harm, being probably of a mechanical or financial nature, is altogether less fearsome than the harm that may ensue if a vehicle is given, by some system of forced induction, a higher performance capability than it was originally designed to exploit. To make a vehicle faster without corresponding attention to the means necessary for making it controllable and roadworthy at its newly elevated speeds is downright irresponsible: the brakes, the suspension, the steering and aerodynamics and tyres, all these, and numerous other parts of the car, must be examined, and if necessary improved to the point where their strength and suitability for high-performance motoring is at least equal to anything that they might have to undertake as a result of the engine's uprating.

Shutdown opens out. Mike Hall with yet another version of the basic drag racing apparel on both sides of the Atlantic. Nitromethane is used up at an alarming rate